High-Linearity CMOS RF Front-End Circuits

Yongwang Ding
Ramesh Harjani

High-Linearity CMOS RF Front-End Circuits

Library of Congress Cataloging-in-Publication Data

A C.I.P. Catalogue record for this book is available
from the Library of Congress.

ISBN 978-1-4419-3663-9
e-ISBN 978-0-387-23802-9

Printed on acid-free paper.

Printed in the United States of America.

9 8 7 6 5 4 3 2 1

springeronline.com

To my parents

Contents

List of Figures

List of Tables

Chapter 1

INTRODUCTION

This book focuses on high performance radio frequency integrated circuits (RF IC) design in CMOS.

1. Development of radio frequency ICs

Wireless communications has been advancing rapidly in the past two decades. Many high performance systems have been developed, such as cellular systems (AMPS, GSM, TDMA, CDMA, W-CDMA, etc.), GPS system (global positioning system) and WLAN (wireless local area network) systems. The rapid growth of VLSI technology in both digital circuits and analog circuits provides benefits for wireless communication systems. Twenty years ago not many people could imagine millions of transistors in a single chip or a complete radio for size of a penny. Now not only complete radios have been put in a single chip, but also more and more functions have been realized by a single chip and at a much lower price.

A radio transmits and receives electro-magnetic signals through the air. The signals are usually transmitted on high frequency carriers. For example, a typical voice signal requires only 30 Kilohertz bandwidth. When it is transmitted by a FM radio station, it is often carried by a frequency in the range of tens of megahertz to hundreds of megahertz. Usually a radio is categorized by its carrier frequency, such as 900 MHz radio or 5 GHz radio. In general, the higher the carrier frequency, the better the directivity, but the more difficult the radio design. There are many radio standards that are employed these days, such as 100 MHz FM radios, 1 GHz analog cell phone radios, 2 GHz digital cell phone radios, and 5 GHz radios in wireless LAN.

In the past radios were built using discrete components: transistors, diodes, inductors, resistors, capacitors, etc. With the development of high speed VLSI process radios can now be made of individual chip sets and each individual chip is built as a stand-alone block with different functions. Many design techniques have also been developed for integrated circuits so that more and more individual blocks are integrated together and a single chip radio is finally becoming a reality.

Besides the integration of more analog circuits into a single chip, the integration of analog circuits into low cost digital CMOS processes also drives the development of modern RF IC designs. With the many new process techniques that has been developed in CMOS in recent years it has now become possible to use CMOS radio chips in high performance systems, such as wireless LAN system and 3G cellular phones.

2. Challenges of modern RF IC design in CMOS

Many modern communication systems are required to handle both very small signals and very large signals with high precision. This in turn requires a low phase noise PLL and a large dynamic range RF front-end, i.e., low noise figure (NF) and high linearity. Many papers and books have discussed the design of a low phase noise PLL in CMOS. This book focuses on the design of a large dynamic range RF front-end.

Although CMOS IC is preferable for its low cost and better integration with DSP chips, it has limitations in term of noise and linearity in comparison to other processes, such as SiGe and GaAs processes. Those limitations place more challenge to designers.

2.1 Noise

The sensitivity of a receiver is defined as the lowest received input power with sufficient SNR at the output. The noise figure of a receiver determines its sensitivity. The lower the noise figure, the better the sensitivity. Equation 1.1 gives the relation between the sensitivitysensitivity and the noise figure, where -174 dBm is the ambient noise floor at 27 C, SNR_{min} is the minimum required signal-to-noise ratio for a certain accuracy and BW is the signal bandwidth.

$$Sensitivity = -174dBm + NF + SNR_{min} + BW \tag{1.1}$$

A noise model of a MOSFET is shown in Figure 1.1 [5]. There are three different noises sources in a MOSFET. i_{nd}^2 is the thermal noise generated by the channel, which is given by Equation 1.2, where k is Boltzmann's constant (1.38 x 10^{-23} JK^{-1}), T is the temperature in Kelvins and g_m is the MOSFET's transconductance. The channel thermal noise is a spectrally white noise.

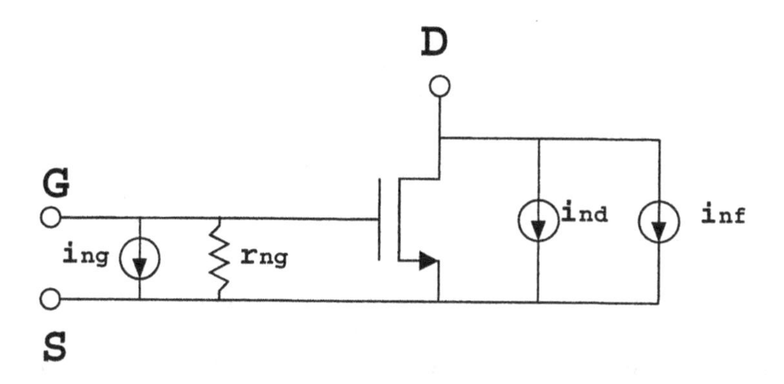

Figure 1.1. Small-signal noise model of a MOSFET

$$i_{nd}^2(f) = 4kT \cdot \frac{2}{3} \cdot g_m \tag{1.2}$$

The flicker noise, as is expressed by v_{nf}^2, is caused by the traps at the interface of the gate oxide and the channel, which is given by Equation 1.3, where K_f is a constant and is dependent on the fabrication process, W and L are the channel width and length, respectively, and C_{ox} is the gate capacitance per unit area. The flicker noise power is inversely proportional to frequency.

$$v_{nf}^2(f) = \frac{K_f}{WLC_{ox}f} \tag{1.3}$$

The induced gate noise, as is expressed by i_{ng}^2, is related to the channel thermal noise and the gate AC coupled current, which is given by Equation 1.5 [49], where δ is the gate noise coefficient, which is equal to $\frac{4}{3}$ for a long channel MOSFET and could be as large as 4 for a very short channel device, and g_{ng} is given by Equation 1.4, where C_{gs} is the gate capacitance, which equals $\frac{2}{3}WLC_{ox}$ for a device in saturation, and g_{d0} is the drain-source conductance at zero drain-source voltage.

$$g_{ng} = \frac{\omega^2 C_{gs}^2}{5g_{d0}} \tag{1.4}$$

$$i_{ng}^2(f) = 4kT\delta g_{ng} \tag{1.5}$$

For most RF front-end circuits operating at radio frequencies both the thermal noise and the induced gate noise dominate the noise figure, and the flicker noise is usually negligible. However, in a mixer where both low frequency and high frequency signals are of concern, the channel thermal noise, the flicker noise and the induced gate noise are all important. Detailed discussions of the noise figure are covered in Chapter 2.

Although a MOSFET creates more noise than a bipolar, it has been shown that a well designed CMOS circuit can provide nearly the same noise figure as bipolar and HBT circuits [17]. Some low noise amplifiers (LNA) have been reported in CMOS, for example, Andrew N. Karanicolas designed a 1.9 dB NF LNA in a 0.5um CMOS [1], Francesco Piazza and Qiuting Huang designed a 1.9 dB NF LNA for their 0.25um CMOS GSM receiver [38], Cheon Soo Kim and et al. designed a 2.8 dB NF LNA in a 0.8um CMOS process [14], and Brian A. Floyd and et al. designed 1.2 dB NF LNA in a 0.8 um CMOS [6].

2.2 Linearity

Linearity is an important parameter that specifies a circuit's ability to handle the large signals. Very large signals can cause desensitization and intermodulation to an imperfectly linear circuit. In general, a MOSFET has a lower linearity than a GaAs device at the same condition. In order to match the performance of GaAs circuits linearization techniques are often used in CMOS IC design.

Desensitization

The 1 dB compression point is the parameter that describes a circuit's tolerance to desensitization. It is defined as the point where the fundamental output is 1 dB lower than what is extrapolated from the linear curve from very small inputs, as is shown in Figure 1.2. A mathematical definition is also given below.

A weak nonlinear circuit with small inputs can be described by Equation 1.6, where x is the input, y is the output, A is the linear gain, α_n is the nth power order nonlinear coefficient, and C is the offset.

$$y(x) = Ax(1 + \alpha_2 x + \alpha_3 x^2 + ...) + C \tag{1.6}$$

If a differential design is used, all the even order products, $\alpha_{2n}x^{2n}$, are cancelled including the DC offset term C. The 1 dB compression point can be calculated by solving Equation 1.7

$$dB[y(x)] = dB[Ax] - 1dB \tag{1.7}$$

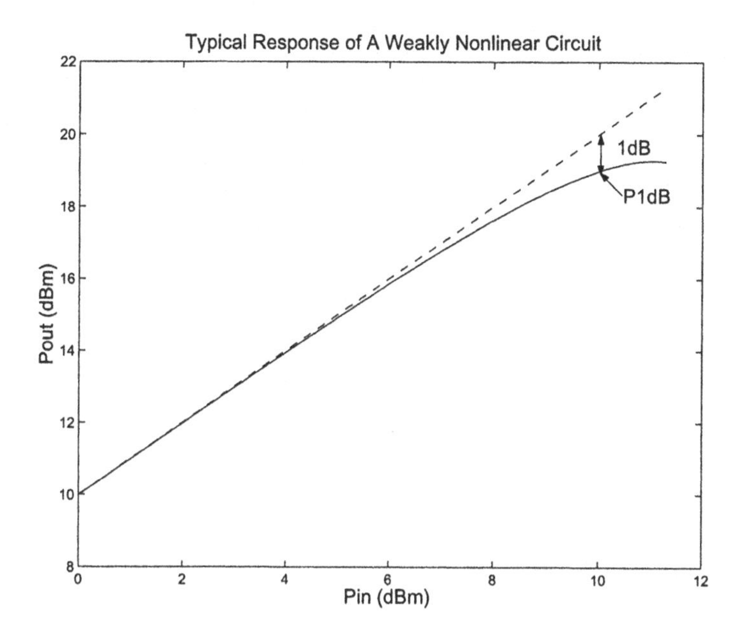

Figure 1.2. Definition of P1dB

P1dB is the 1dB compression point in power at either the input or the output. It is one of the key specifications of power amplifiers. Because MOSFETs have a lower breakdown voltage and higher knee voltage than bipolars, they have a lower P1dB, which in general limits the output power for linear operation. Some techniques have been developed to improve the P1dB of a CMOS power amplifiers. Power amplifier details will be discussed in Chapter 7.

Intermodulation

The intercept point is another important parameter that describes the circuit's linearity and is often used to indicate the level of intermodulation. N^{th} order intercept point is defined as the cross point where the linear extrapolation of the fundamental signal equals the linear extrapolation of the nth order harmonic. An example of the 3rd order intercept point is shown in Figure 1.3. If there is only 3rd order distortion in the circuit, it is easy to show that IP3 (3rd intercept point) is 9.6 dB higher than the 1 dB compression point, which is also expressed in Equation 1.8 [10].

$$P_{IP3} = P_{1dB} + 9.6dB \tag{1.8}$$

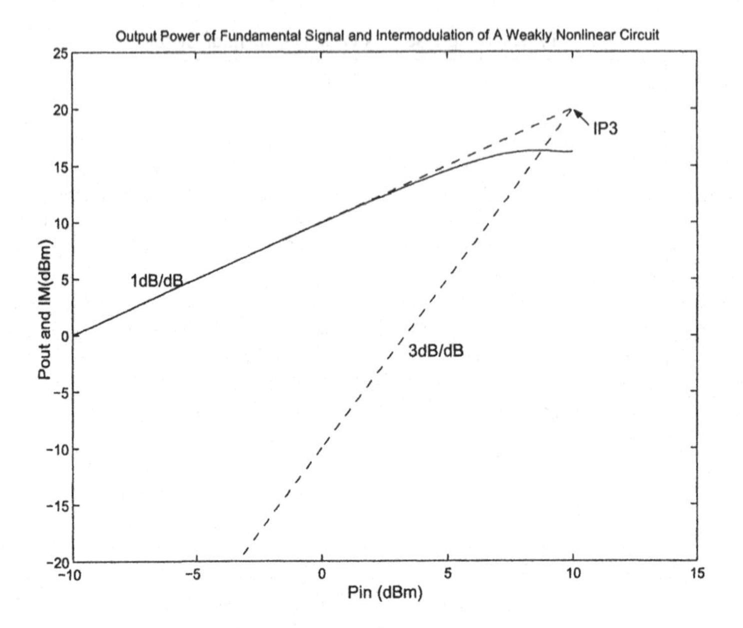

Figure 1.3. Definition of IP3

IP3 is often specified in a receiver to describe its ability to handle large signals. The higher the IP3, the smaller the intermodulation signal. The dynamic range is defined as the range between the largest signal to produce sufficiently low intermodulation and the sensitivity. Equation 1.9 gives a relation between dynamic range, IP3 and sensitivity.

$$Dynamic\, range(dB) = IIP3 - \frac{SNR_{min}}{2} - sensitivity \tag{1.9}$$

Because of mobility degradation, channel length modulation and other imperfect in its I-V characteristic a MOSFET has a lower IP3 than a GaAs device. Several techniques have been developed to improve the IP3 of a MOSFET. The details are discussed in Chapters 3 and 4.

3. Contributions of this work

In order to increase the performance of a radio both the dynamic range of the receiver and the output power of the transmitter need to be maximized. The dynamic range of the receiver is determined by the sensitivity and the linearity of its RF front-end; and the output power of the transmitter is usually limited by the linearity of its power amplifier. Therefore maximizing the linearity of the RF circuits becomes very critical to radio design.

Many techniques have been developed to linearize integrated circuits [54, 19, 24, 34, 18, 2, 16], but most of them are limited to low frequency applications and are not suitable for giga-Hertz radios, as will be discussed in Chapter 3. A new linearization technique is presented in Chapter 4 to improve the linearity of integrated circuits in general. With a 1% process matching the output linearity can be improved by up to 40 dB. No prior knowledge of distortion or linear characteristic of the circuits is required and there is only a slight trade-off of power consumption, gain and noise.

A prototype low noise amplifier (LNA) has been designed and fabricated in the TSMC 0.25 μm CMOS process. As is shown in Chapter 5, the LNA has been tested with a 15 dB gain, 2.8 dB NF and +18 dBm IIP3. The linearity in term of IIP3 is improved by 13 dB with the help of the new linearization technique. Furthermore, a receiver RF front-end, composed of a LNA and a down conversion mixer, is designed with the same technique in Chapter 6; and the results show a +12 dB increase in dynamic range with only 12% more DC power consumption.

Additional techniques are presented in Chapter 7 to increase the output power of the transmitter. A test chip fabricated in the UMC 0.18 μm CMOS process shows a single MOSFET amplifier can improve the output referred P1dB by 3 dB. A new parallel class A&B power amplifier is designed in CMOS. It not only increases the output power by more than 3 dB but also reduces the DC power consumption by over 50% compared to a class A CMOS amplifier. Measurement results show the class A&B amplifier has 12 dB gain, +22 dBm output power and more than 44% PAE.

Chapter 2

RF DEVICES IN CMOS PROCESS

1. Introduction

Circuits are made of individual components, such as transistors, inductors, capacitors, resistors, etc. A good circuit design has to make full use of each component. That requires a thorough understanding of the characteristic of each device.

This chapter discusses individual devices in CMOS ICs, such as MOSFETs, spiral inductors, metal capacitors, thin film resistors, etc.

2. MOSFET

The MOSFET is the core device in CMOS ICs. Based on the channel doping it is categorized into two types, NMOS and PMOS. An NMOS has n type doped diffusion and p type doped substrate; likewise, a PMOS has p type doped diffusion and n type doped substrate. Figure 2.1 shows a cross section of a typical NMOS. Because the performance of the MOSFET determines the quality of CMOS circuits, it is very important to understand the characteristics of a MOSFET, such as transconductance, noise figure and linearity.

2.1 Transconductance

MOSFETs are often used to translate an input voltage into an output current. Transconductance is the ratio of the output current to the input voltage. With an appropriate impedance at the output a MOSFET can provide voltage gain; likewise, with an appropriate conductance at the input a MOSFET can provide current gain.

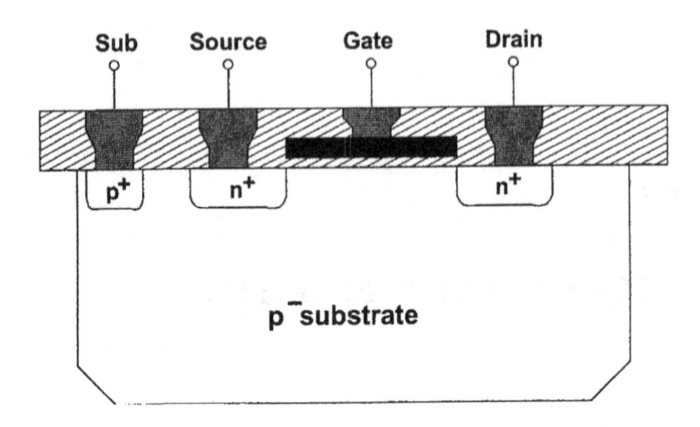

Figure 2.1. Cross section of an NMOS

There are two operating modes of a MOSFET. When the gate-to-source voltage V_{gs} is smaller than the threshold voltage V_{th}, the channel is in accumulation mode, i.e., the channel is off. When V_{gs} is larger than V_{th}, the channel is in inversion mode, i.e., the channel is on.

However, in actual operation there is no distinct line between on and off. When V_{gs} is equal to or even slightly smaller than V_{th}, there is still a small amount of current flowing from drain to source, which is called the subthreshold current. This intermediate region is called either subthreshold region or weak inversion region. The drain current in subthreshold region is given by Equation 2.1 [19], where I_d is the drain current, W and L are the gate width and length, respectively, V_q equals kT/q, k is Boltzmann's constant (1.38 x 10^{-23} J/K), T is the temperature in Kelvins, n is a constant usually around 1.5, and I_{d0} is another constant around 20 nA.

$$I_d = I_{d0}\left(\frac{W}{L}\right)e^{\frac{V_{gs}}{nV_q}} \tag{2.1}$$

A typical NMOS I-V curve is shown in Figure 2.2. The two operating regions are defined as triode region and saturation region. In the triode region V_{gs} is larger than V_{th} and the drain-to-source voltage V_{ds} is smaller than the over-drive voltage V_{od}, which equals to $V_{gs} - V_{th}$. In this region the drain current is a strong function of V_{ds}, as is given by Equation 2.2, where μ_n is the mobility and C_{ox} is the gate capacitance per unit area.

$$I_d = \mu_n C_{ox}\frac{W}{L}\left[\left(V_{gs} - V_{th}\right)V_{ds} - \frac{V_{ds}^2}{2}\right] \tag{2.2}$$

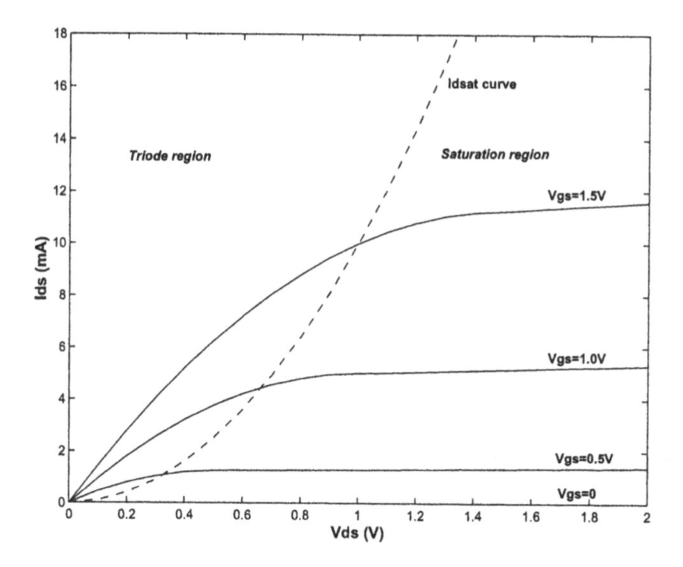

Figure 2.2. I-V curve of a typical NMOS

When V_{ds} is much smaller than V_{od}, the drain current is almost proportional to V_{ds}, as is given in Equation 2.3. A transistor with such a bias can be used as an effective resistor. The resistance is given by Equation 2.4.

$$I_d \mid_{V_{ds} \ll V_{od}} \simeq \mu_n C_{ox} \frac{W}{L} \left(V_{gs} - V_{th}\right) V_{ds} \tag{2.3}$$

$$R_{on} \mid_{V_{ds} \ll V_{od}} \simeq \frac{1}{\mu_n C_{ox} \frac{W}{L} \left(V_{gs} - V_{th}\right)} \tag{2.4}$$

When V_{ds} is larger than V_{od}, a MOSFET is in saturation region. A general relation of the output current versus the input voltage of an MOSFET in saturation is given by Equation 2.5 [49], where μ_0 is the mobility when the over-drive voltage is zero, v_{sat} is the channel carrier saturation velocity extrapolated from the low electrical field, α is given by Equation 2.6, and θ is a modification factor of the mobility as shown in Equation 2.7.

$$I_d = \frac{1}{2} \mu_0 C_{ox} \frac{W}{L} \frac{V_{od}^2}{1 + \alpha V_{od}} \tag{2.5}$$

$$\alpha = \theta + \frac{\mu_0}{2 v_{sat} L} \tag{2.6}$$

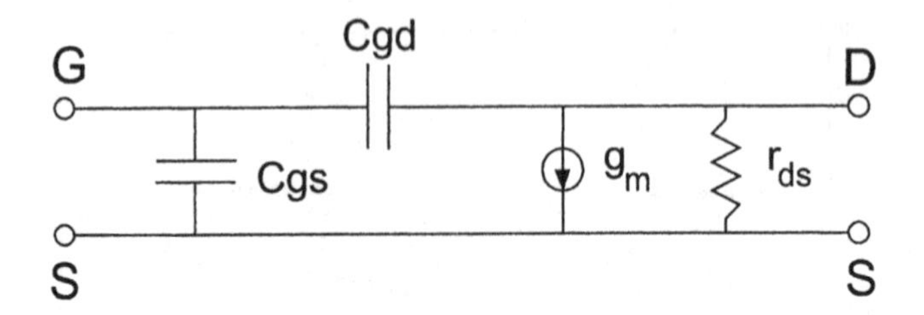

Figure 2.3. Small-signal model of a MOSFET in saturation

$$\mu_n = \frac{\mu_0}{1 + \theta V_{od}} \tag{2.7}$$

The small-signal transconductance of a MOSFET in saturation is given by Equation 2.8.

$$g_m = \mu_0 C_{ox} \frac{W}{L} V_{od} \frac{1 + \frac{\alpha}{2} V_{od}}{(1 + \alpha V_{od})^2} \tag{2.8}$$

When the channel length is long, i.e., $V_{od} \ll LE_{sat}$, the drain current is approximated by Equation 2.9 and the small-signal transconductance is approximated by Equation 2.10.

$$I_d = \frac{\mu_n C_{ox}}{2} \frac{W}{L} V_{od}^2 \tag{2.9}$$

$$g_m = \mu_n C_{ox} \frac{W}{L} V_{od} \tag{2.10}$$

2.2 Small-signal model

The small-signal model of a MOSFET is very useful to circuit design. Figure 2.3 shows a simplified model of a MOSFET in saturation. Here, g_m is the small-signal transconductance, C_{gs} is the gate-to-source capacitance, C_{gd} is the overlap capacitor between the gate and the drain, and r_{ds} is the drain impedance.

This model is used for low frequencies. However, it is not sufficient to represent a MOSFET at giga-Hertz frequencies. Some parasitic components related to the physical layout of the MOSFET need to be considered at high frequencies as well. A RF small-signal model is shown in Figure 2.4. R_g is

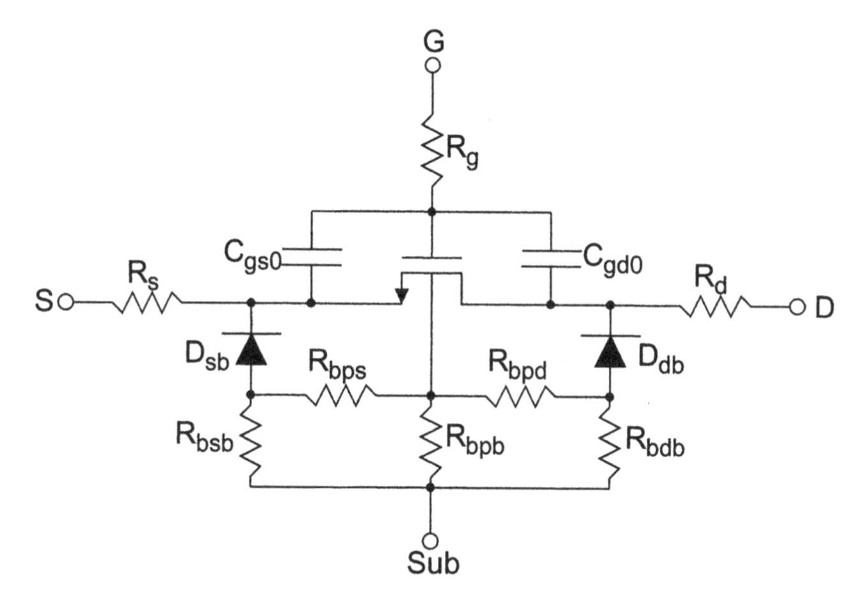

Figure 2.4. RF small-signal model of a MOSFET

the gate parasitic resistance, which usually consists of the distributed poly gate resistance and the parasitic resistance of the routing metal wires. R_s and R_d are the parasitic resistance at the source and at the drain, respectively. D_{sb} and D_{db} are the junction diodes at the source and at the drain, respectively. C_{gs0} is the over-lap capacitor between the gate and the source, and C_{gd0} is the over-lap capacitor between the gate and the drain. R_{bps}, R_{bpd}, R_{bsb}, R_{bpb} and R_{bdb} represent the substrate resistive network, which connects every point in the substrate, such as the points close to the source, to the drain and to the channel, respectively, and the substrate contact point.

Figure 2.5 shows a simple amplifier and Figure 2.6 gives the simulation results with different models. The example uses an NMOS with 128 μm width and 0.18 μm length, and an LCR tank at the output in the UMC18 μm RF CMOS process. The simulations show more than 1 dB degradation in gain at the resonant frequency with the RF model than that with the base model (BSIM3).

2.3 Linearity

A high dynamic range system requires high linearity. The transconductance of a perfectly linear MOSFET should be constant over the input voltage. The

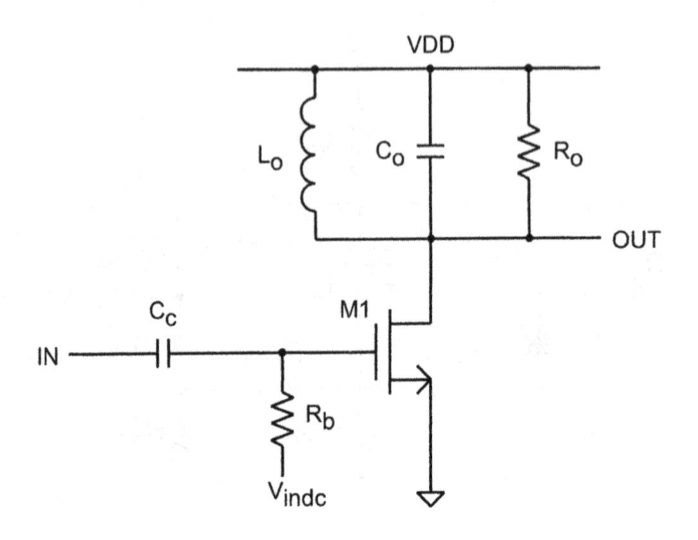

Figure 2.5. Simple amplifier

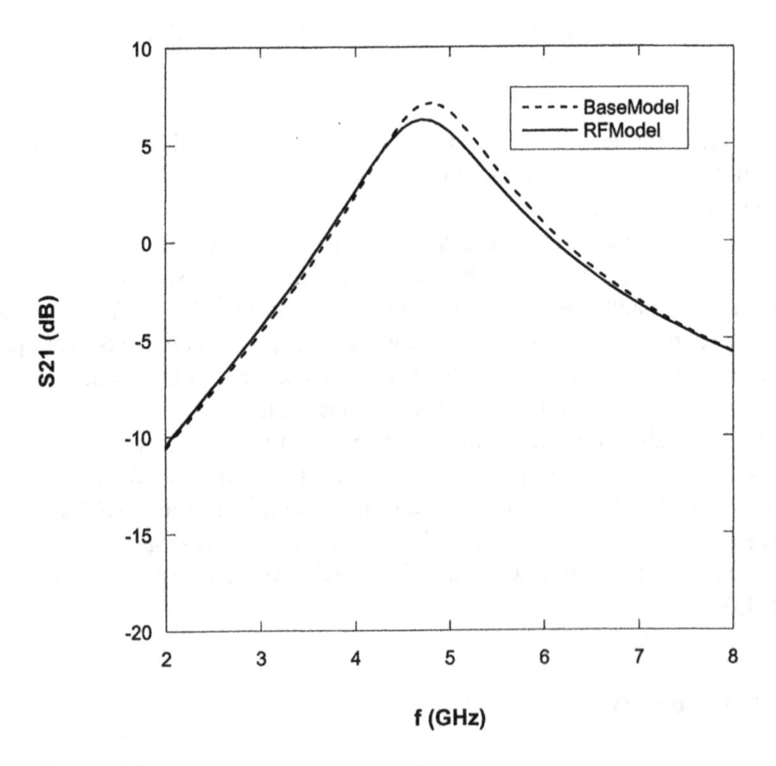

Figure 2.6. Simulated gain of the simple amplifier (Figure 2.5)

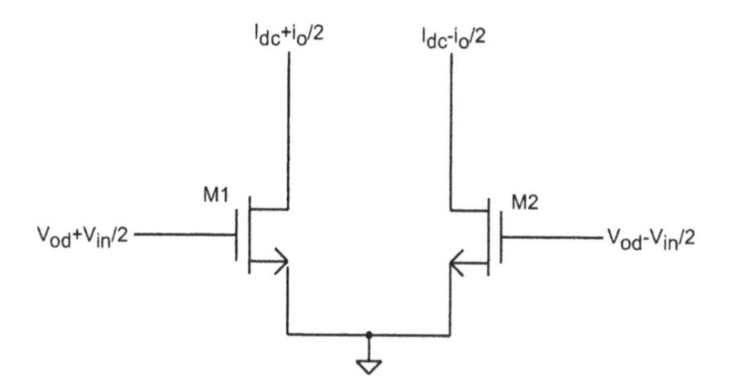

Figure 2.7. Common-source grounded MOSFET pair

transconductance dependence on the input voltage indicates the nonlinear performance of MOSFETs.

As shown in Equation 2.10, only second order distortion is generated by a long channel MOSFET. If a differential design is applied, this second order distortion will be cancelled at the output, as is given by Equation 2.11. A long channel common-source grounded MOSFET pair in Figure 2.7 is a linear transconductor cell.

$$\begin{aligned} i_+ &= \mu_n C_{ox}\frac{W}{L}(V_{od}+\frac{V_{in}}{2})^2 \\ i_- &= \mu_n C_{ox}\frac{W}{L}(V_{od}-\frac{V_{in}}{2})^2 \\ i_{out} &= i_+ - i_- \\ &= \mu_n C_{ox}\frac{W}{L}V_{od}V_{in} \end{aligned} \tag{2.11}$$

Short channel MOSFETs generate odd order distortion at the output even with a differential design. Usually the third order distortion dominates when the input signal is small. The input referred 3rd intercept voltage point (IIV3) of a MOSFET is given by Equation 2.12 [48].

$$IIV_3 = \sqrt{\frac{8}{3}\frac{1}{\alpha}V_{od}\left(1+\frac{1}{2}\alpha V_{od}\right)(1+\alpha V_{od})^2} \tag{2.12}$$

Note that the IIV_3 increases with the over-drive voltage V_{od}. However, the DC power consumption also increases with V_{od}. There is always a trade off between the linearity and the power consumption in linear circuit design.

As shown in Equation 2.12 the channel length is an important factor in regards to linearity. The longer the channel, the higher the linearity, but the slower the

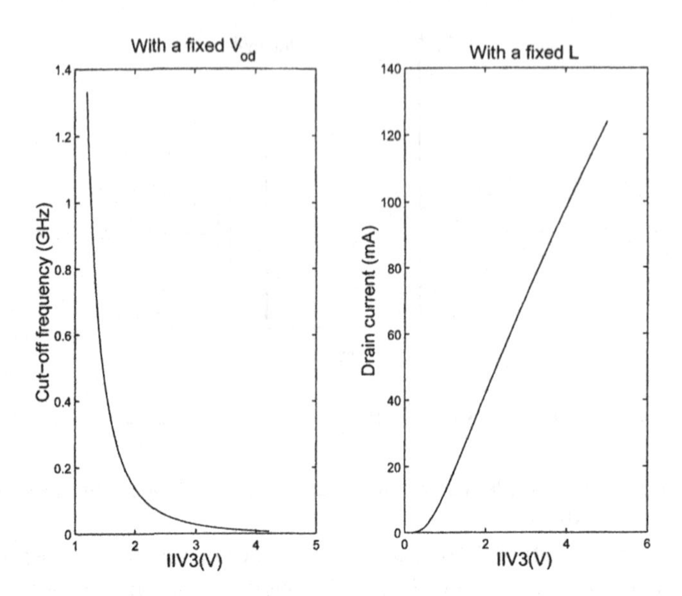

Figure 2.8. IIV3 vs. drain current and cut-off frequency of a typical NMOS

circuit's response. The cut-off frequency, ω_T, as defined in Equation 2.13, is almost inversely proportional to the square of the channel length. Therefore, linear circuit design is also a balance between the linearity and the speed.

$$\omega_T = \frac{g_m}{C_{gs}} \approx \frac{3}{2}\frac{\mu_n V_{od}}{L^2} \tag{2.13}$$

Figure 2.8 shows the correlation among the linearity, the drain current and the cut-off frequency of a typical NMOS.

2.4 Noise

There are three major noise contributions in a MOSFET, thermal noise generated in the resistive channel, flicker noise generated by the surface traps between the silicon and the gate oxide and shot noise related to the junction diodes [5].

Both the channel thermal noise and the junction shot noise are often considered as white noise. The channel thermal noise is much more significant for MOSFET than the shot noise. The channel thermal noise of a MOSFET is given by Equation 2.14.

$$i^2_{n,thermal}(f) = \frac{8}{3}kTg_m \tag{2.14}$$

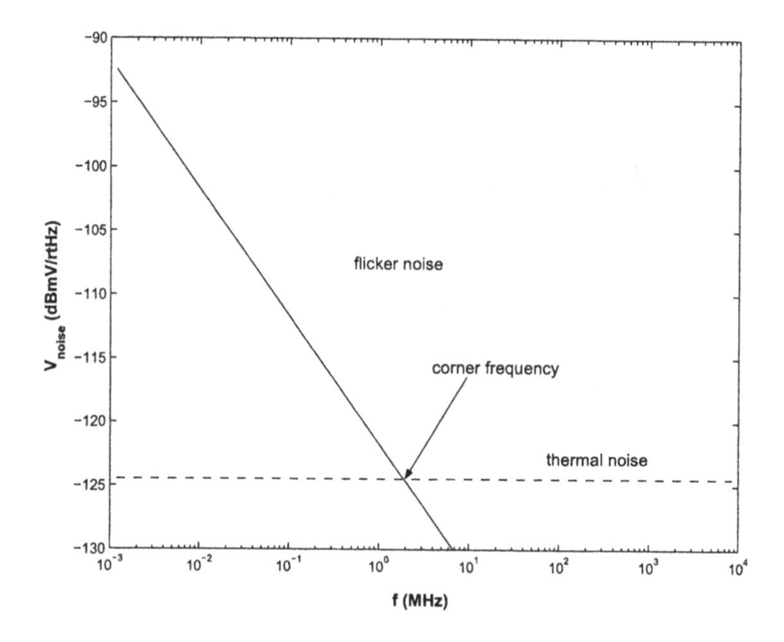

Figure 2.9. Flicker noise corner frequency

The equivalent input referred noise voltage of the channel thermal noise is then given by Equation 2.15.

$$v_{in,thermal}^2(f) = \frac{8}{3}\frac{kT}{g_m} \tag{2.15}$$

Equation 2.16 gives the input referred noise voltage of the flicker noise, where K_f is a process constant. Note that flicker noise is inversely proportional to the frequency and becomes a dominant noise source at very low frequencies. The flicker noise corner frequency is defined as the frequency where the flicker noise equals the channel thermal noise, as is illustrated in Figure 2.9. For the majority of IF circuits it is desirable to use large-sized transistors so that their corner frequencies is lower than the signal band.

$$v_{n,flicker}^2(f) = \frac{K_f}{WLC_{ox}f} \tag{2.16}$$

Another important noise source in MOSFETs is induced gate noise. Induced gate noise is caused by some of the channel thermal noise being coupled into the gate. Equation 2.17 gives the input referred induced gate noise, where δ is the gate noise coefficient, which is $\frac{4}{3}$ for long channel devices and could be

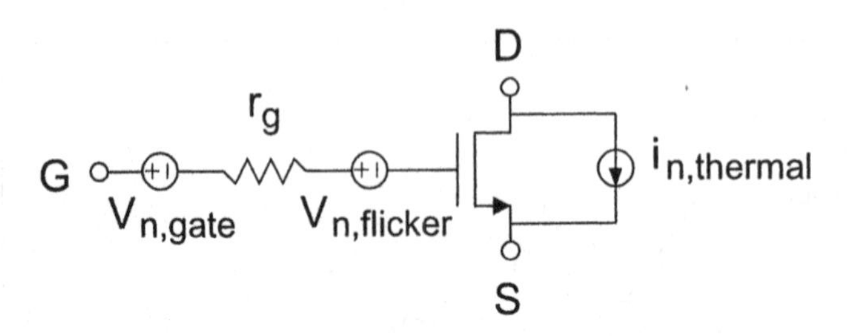

Figure 2.10. Noise model of MOSFET

as large as 4 for very short channel devices [5]. Additionally, r_g is the gate noise resistor and is given by Equation 2.18 as a function of the zero-biased drain conductance g_{d0}. The induced gate noise increases with the frequency; therefore, it is also called blue noise.

$$v_{n,gate}^2(f) = 4kT\delta r_g \tag{2.17}$$

$$r_g = \frac{1}{5g_{d0}} \tag{2.18}$$

Studies have shown that the induced gate noise is correlated to the channel thermal noise. The correlation, which is usually defined as c, is j0.395 for long channel devices.

The complete two port noise model is shown in Figure 2.10. Because the flicker noise and the induced gate noise dominate at different frequency ranges, simplified noise models can be used accordingly. Figure 2.11 shows the simplified noise model for low frequencies, and Figure 2.12 shows the simplified noise model for high frequencies.

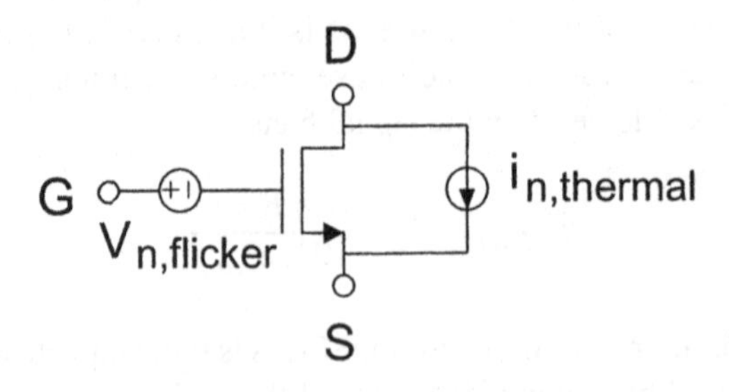

Figure 2.11. Noise model of MOSFET at low frequencies

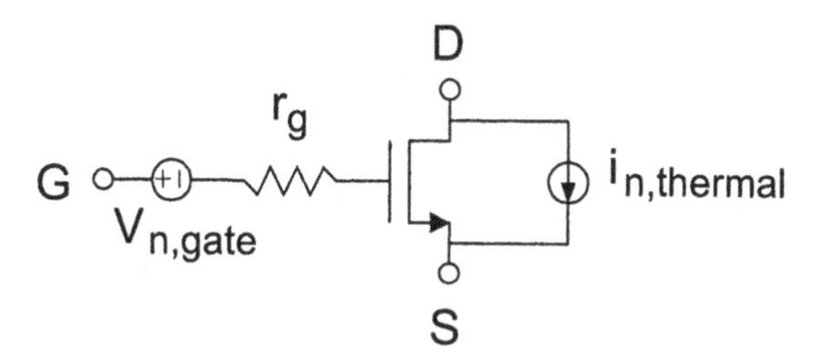

Figure 2.12. Noise model of MOSFET at high frequencies

3. Inductor

The inductor is one of the most critical passive devices in RF ICs. Inductors are used to provide filtering, increase isolation, compensate parasitic capacitors, etc. They have a significant impact on both the performance and the cost of the complete system. In order to provide high integration to lower the cost of IC products, on chip inductors are always preferable to off chip components. However, on chip inductors are also most area-consuming and are difficult to design in CMOS.

Spiral planar inductor is one popular on-chip inductor in CMOS. It has benefits of easy layout and relative insensitiveness to process. If a thick metal layer is available for the top layer winding, the spiral inductor can have higher quality factor (Q). For example, in a typical pure digital CMOS process a spiral planar inductor with a 0.6 μm thick aluminum layer can provide a Q of 4 to 6, and in a typical RF CMOS process a spiral planar inductor with a 2.0 μm thick aluminum layer can easily provide Q that is greater than 10.

3.1 Layout

The on chip inductors have several different shapes, such as square, octagon and circle. Square spiral inductor is easiest for layout and also most developed in terms of modeling and simulation. However, it is also the least area efficient and results in the lowest quality. Figure 2.13 shows an example of a square inductor.

Circular spiral inductors are the most area efficient and provide the highest quality. However, it is not available in many CMOS processes because it requires infinite fine angle step. Figure 2.14 shows an example of a circular inductor.

Octagonal inductors have benefits of area efficiency and quality, and compatibility to most CMOS processes. Therefore, they are the most preferred inductors on chip. A layout example is show in Figure 2.15.

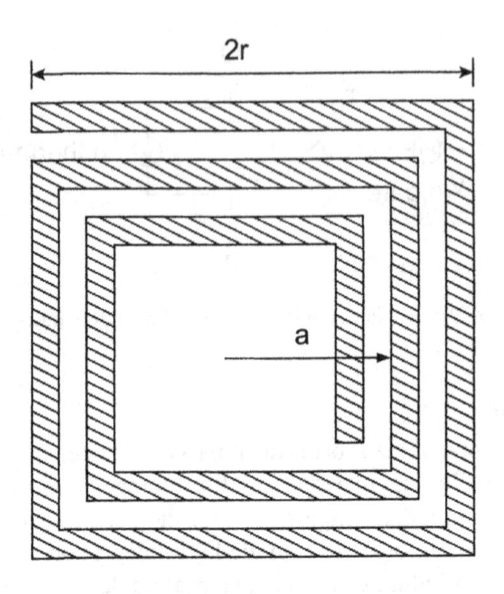

Figure 2.13. Square spiral inductor

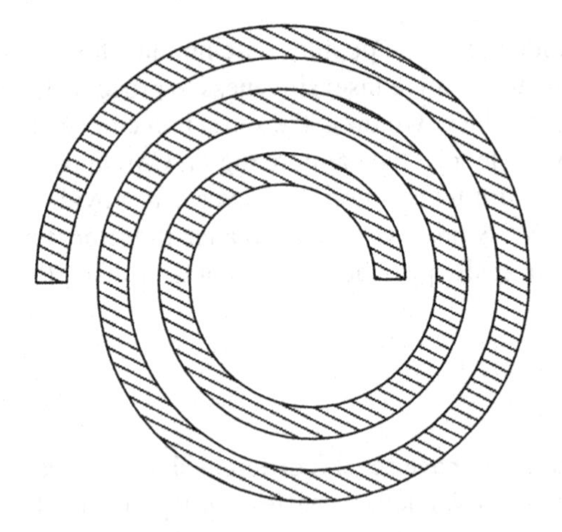

Figure 2.14. Circular spiral inductor

The substrate is another important factor to the performance of an on-chip spiral inductor. When an AC current flows through the inductor a fluctuating magnetic flux is created, which in turn generates an eddy current in the substrate if the substrate is conductive. Such a current not only decreases the total effective inductance, but also degrades the quality of the inductor because it effectively couples the substrate resistance into the inductor. In order to reduce or eliminate the eddy current, either a patterned ground shield or a highly resis-

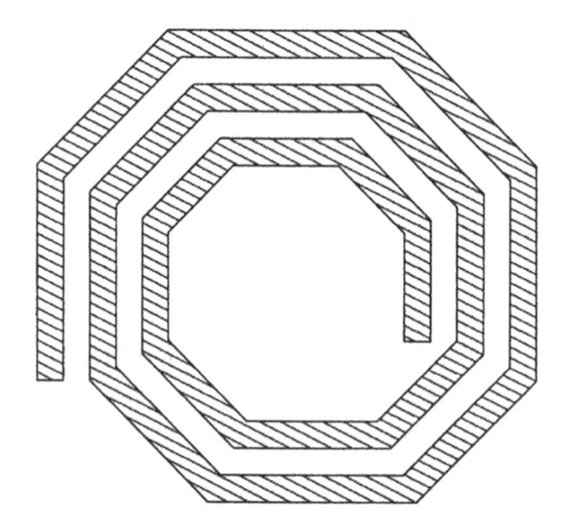

Figure 2.15. Octagonal spiral inductor

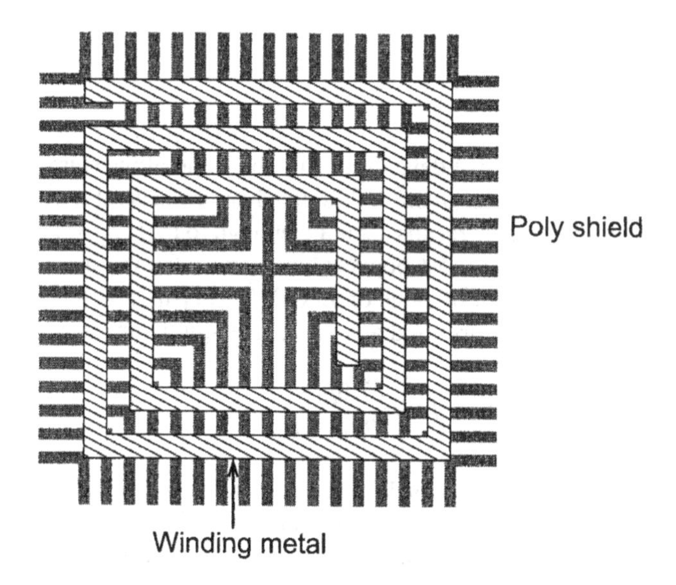

Figure 2.16. Square spiral inductor with polysilicon patterned ground shield

tive substrate is used [13, 12]. Figure 2.16 shows an example of a square spiral inductor with a polysilicon patterned ground shield. The ground shield is laid out perpendicular to the current direction so that no eddy current is able to be generated. The shield also has to be grounded very well to stop any electrical field from penetrating to the substrate.

Another option to prevent eddy current is to create patterned deep trenches in the substrate. The trenches are deeply oxidized so that they are non-conductive.

Layout	K_1	K_2
Square	2.34	2.75
Hexagonal	2.33	3.82
Octagonal	2.25	3.55

Table 2.1. Coefficients of modified Wheerler's formula

The substrate has to be highly conductive and connected to ground in order to prevent the penetration of the electrical field. The pattern of the deep substrate trenches is the same as the polysilicon patterned ground shield. However, deep trench substrates are not available in most of CMOS processes.

In general an octagonal spiral inductor with patterned ground shield or highly resistive substrate is often preferable in CMOS process for its area efficiency and quality factor.

3.2 Simulation

Compared to other passive devices on chip the value of the spiral inductor is difficult to calculate from its layout. Although there are several formulas to use, none of them is able to give results accurate enough to meet most RF IC design requirements. One of the most popular equation is shown in Equation 2.19 [28], where μ_0 is a constant of value of 4πx10^{-7} henry/m, r is the radius of the spiral, n is the number of turns and a is the square spiral's mean radius.

$$L = \frac{37.5\mu_0 n^2 a^2}{22r - 14a} \tag{2.19}$$

More accurate formula based on Wheeler's Equation is given in Equation 2.20 [46], where d_{avg} is the average diameter and equals $2a$; ρ is the fill ratio, and equals $(r - a)/(r + a)$; and K_1 and K_2 are the layout dependent coefficients and are given in Table 2.1.

$$L = K_1 \mu_0 \frac{n^2 d_{avg}}{1 + K_2 \rho} \tag{2.20}$$

In order to achieve more accurate spiral inductor results, some complicated simulation tools have been developed. The most accurate simulation tools are 3D electro-magnetic (EM) simulators, such as HFSS and Sonnet. The 3D EM simulators usually require heavy simulation power and long simulation time. Some compromised tools between 3D EM simulation and empirical finite

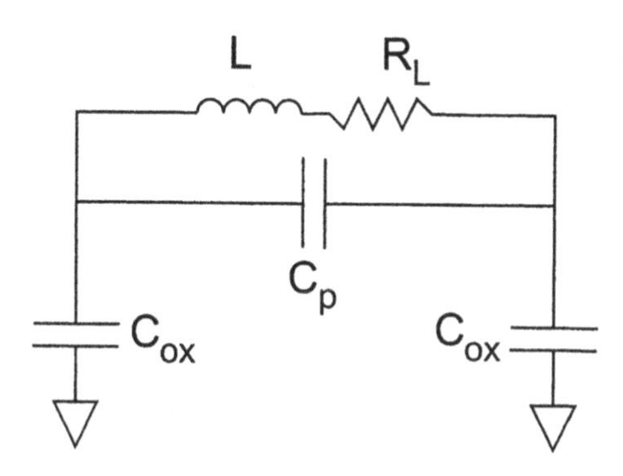

Figure 2.17. 5-element spiral inductor model

element approximation were also created, such as ASITIC and OEA SPIRAL. The compromised simulators provide a good balance between accuracy and simulation complexity.

3.3 Modeling

A practical on-chip spiral inductor is not a pure inductive component, but a complex structure with inductor, parasitic resistors and parasitic capacitors. It is important for designers to have a lumped model to visualize the limitation of the spiral inductor and optimize the circuit design.

Figure 2.17 shows a simple 5-element inductor model. This model assumes a perfect patterned ground shield so that there is no eddy current or substrate-coupled resistance. If an imperfect ground shield is used or a low conductive substrate is used, substrate-coupled resistors (R_{sub}) have to be included. Figure 2.18 shows a 9-element inductor model with substrate resistors and capacitors.

On chip spiral inductors always have a significant finite series resistance, which limits the quality of the inductors. At low frequencies the current flowing through the winding metal is almost evenly spread; therefore, the series resistance is only a function of the winding metal width and total length. At high frequencies, especially at giga hertz, the AC current starts to crowd towards the edge of the metal; therefore, besides the metal width and length, the series resistance is also a function of the operating frequency and the metal thickness. This is called skin effect. In order to include the skin effect, an 11-element model is used as shown in Figure 2.19 [8].

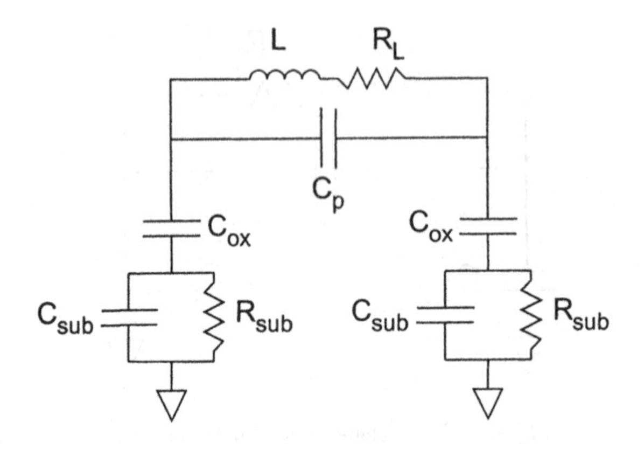

Figure 2.18. 9-element spiral inductor model

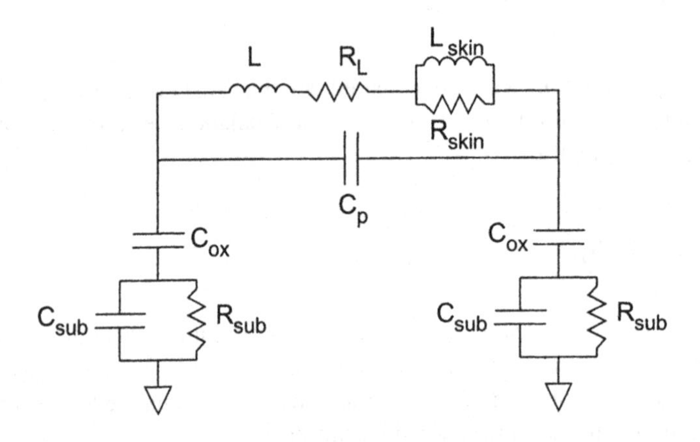

Figure 2.19. 11-element spiral inductor model

Although the 11-element model is more accurate to characterize an on-chip inductor, it requires more complicated extraction. For most narrow band circuits 9-element model is sufficient and is easier to extract.

4. Capacitor

The capacitor is another important passive device in analog circuits. It is used for AC coupling, filtering, de-coupling for supply voltages, etc. There are different types of capacitors available in CMOS processes. Some are linear but with smaller value per area, such as metal capacitors and polysilicon capacitors; and some are area efficient but have a nonlinear characteristic over voltage, such as MOS caps and varactors.

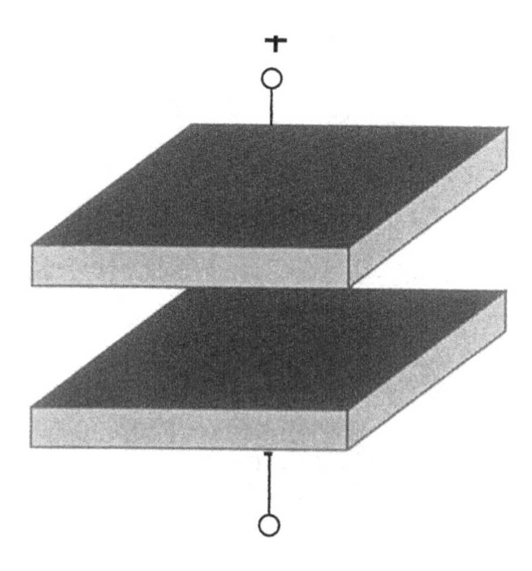

Figure 2.20. Metal plate capacitor

4.1 Metal capacitor

A metal capacitor is made of two pieces of metal in parallel. Generally the value of a metal capacitor is determined by the distance of the two pieces of metal and the dielectric constant of the material between them. In CMOS the dielectric material is silicon dioxide. Because the silicon dioxide doesn't change either its dielectric constant or its shape over voltage or temperature, the value of a metal capacitor is nearly constant over voltage and temperature.

A plate capacitor is made of two layers of metal, as is shown in Figure 2.20. In a pure digital CMOS process, the capacitance per unit area of two layers of metal is usually very small and the parasitic capacitance of the bottom plate is comparable to the main capacitance if only two layers of metal are used. One way to increase the capacitance per unit area is to use a sandwich structure with multiple layers of metal. Figure 2.21 shows an example of a sandwich metal capacitor.

A metal capacitor can also be made of two interdigitated metal in the same layer, as is shown in Figure 2.22. Because the allowable minimum spacing between metal in the same layer is usually smaller than the distance between two layers of metal, and also because the effective area is increased by a number of fingers, the capacitance per unit area of a interdigitated metal capacitor is higher than that of two layers of metal.

Furthermore, the sandwich structure and the interdigitated structure are often used together to build a metal capacitor in a pure digital CMOS process. Not

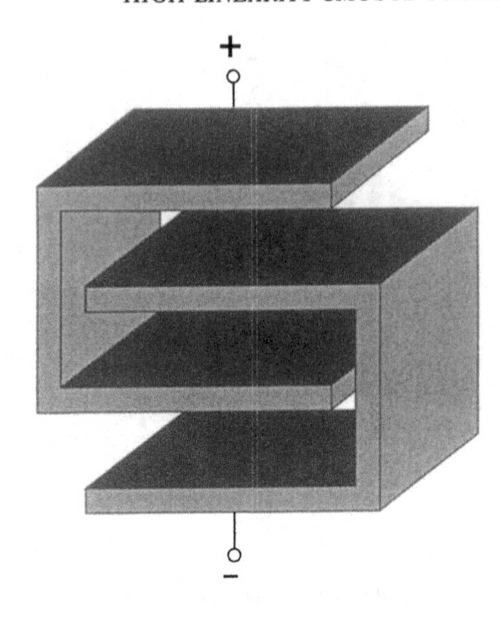

Figure 2.21. Metal sandwich capacitor

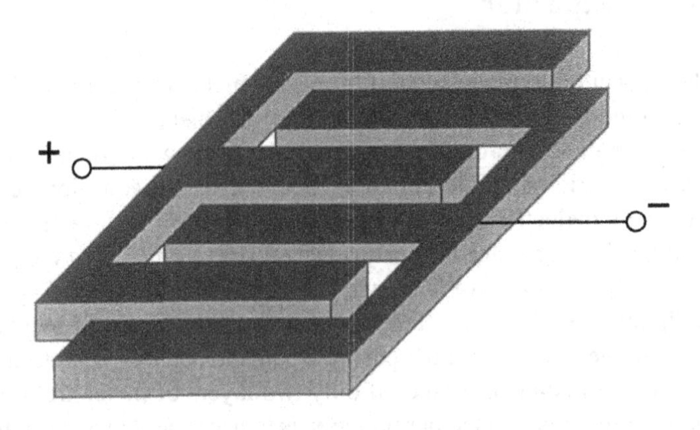

Figure 2.22. Interdigitated metal capacitor

only the capacitance per unit area is increased, but also the ratio of the parasitic capacitance to the main capacitance is reduced.

In some CMOS process, a high value capacitor can be made with help of a special metal layer. Because the distance of the special metal to the second metal layer is 20 to 30 times smaller than the distance between two standard adjacent metal layers, the area capacitance is 20 to 30 times higher than the normal metal plate capacitance. The parasitic capacitance of the bottom plate of this special metal capacitor is much smaller than that of metal plate capacitors. Therefore, it is preferable in RF CMOS ICs.

Metal capacitors have high quality factor (Q) in general. When they are used in parallel with on-chip inductors, their parasitic resistance is often ignored.

4.2 Polysilicon capacitor

A polysilicon capacitor is made of two polysilicon layers. Its structure is very similar to a metal plate capacitor except the distance between the polysilicon layers is much smaller than that between metal layers. Therefore, the capacitance per unit area of a polysilicon capacitor is much higher than that of a metal plate capacitor. However, because polysilicon has a much higher sheet resistance than metal, usually by more than 100 times, the quality factor of a polysilicon capacitor is much lower. Also the bottom polysilicon layer is so close to the substrate, i.e., typically within a micron of the field oxide, that a polysilicon capacitor has a significant parasitic capacitance on its bottom plate. Therefore, polysilicon capacitors are often used in low frequency circuits but not RF ICs.

4.3 Varactor

A varactor is a variable capacitor which is usaully made of an active device, such as a pn junction diode or a MOSFET. It is mostly used to tune the resonant frequency of a LC tank, such as in VCO, but sometimes is also used to stabilize supply voltages and DC bias voltages.

One of the most important characteristic of a varator is the capacitance dependence on voltage. For example, a pn junction diode has a junction capacitance which is a function of its depletion width, which is in turn a function of the voltage across the junction. Equation 2.21 gives the capacitance dependance of a linearly graded pn junction [42], where V_A is the voltage across the pn junction and C_{j0} is the junction capacitance when V_A equals zero.

$$C_j = \frac{C_{j0}}{\left(1 - \frac{V_A}{V_{bi}}\right)^3} \tag{2.21}$$

A varactor can also be made using a MOSFET. There are two basic types of MOS varactors: the accumulation mode MOS varactor and the inversion mode MOS varactor [42]. The diffusion and the bulk of the accumulation mode MOS varactors are doped to the same polarity, i.e., both are n-type doped or p-

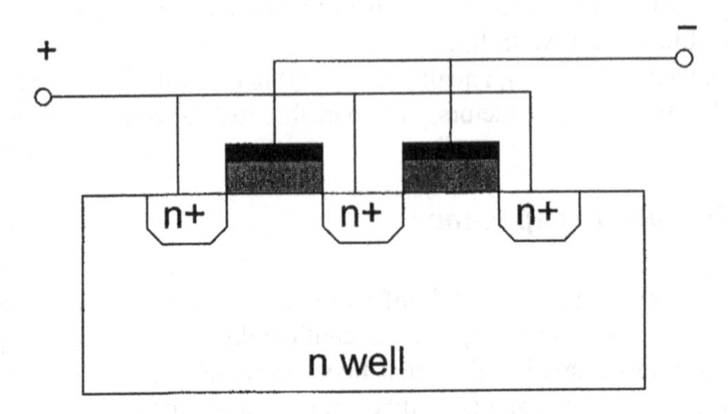

Figure 2.23. Accumulation mode varactor

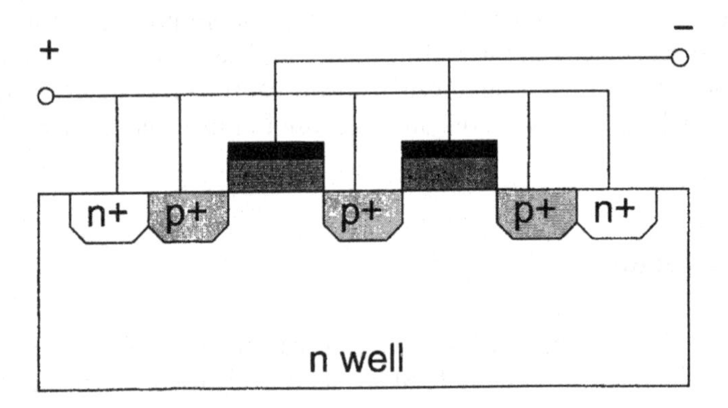

Figure 2.24. Inversion mode varactor

type doped, but with different degrees; therefore, an accumulation layer usually exists at the surface of the channel, and its depth changes with the voltage across the gate and the channel. On the other hand, the diffusion and the bulk of the inversion mode MOS varactor are doped to the different polarity, i.e., one is n-type doped and the other is p-type doped; therefore, an inversion layer can form at the surface of the channel when a voltage is applied to the gate to attract the opposite polarity charge from the channel, and its depth also depends on the voltage across the gate and the channel. Figure 2.23 shows an example of an accumulation mode MOS varactor and Figure 2.24 shows an example of an inversion mode MOS varactor.

Both the MOS varactors have a higher quality factor Q than the pn junction varactor [4]. Also the two types of MOS varactors have much higher capacitance per unit area than the pn junction varactor, which makes the MOS varactors better candidates for de-coupling of supply voltages and DC bias voltages. Between the two types of MOS varactors the accumulation mode varactor has a less sharp slope of tuning capacitance over voltage, which implies less multiplicative phase noise when used in a VCO.

5. Resistor

There are many types of resistors available in CMOS processes. Based on the material used they can be categorized into four groups, metal resistors, polysilicon resistors, diffusion resistors and well resistors. Both metal resistors and polysilicon resistors have low variation with voltage, but diffusion resistors and well resistors have a strong dependence on voltage.

Metal resistors have the lowest sheet resistance, e.g. 10 to 80 milli-ohms per square. Because of such low resistance, metal resistors are usually used to track the metal resistance change over process and temperature.

Well resistors have a large sheet resistance, e.g. 300 to 500 ohms per square. Because of their nonlinear characteristic over voltage, well resistors are usually used to feed a DC bias voltage, and have to be used with great caution so as to not introduce distortion.

Diffusion resistors have a moderate range of sheet resistance, e.g. around 10 ohms per square if salicided and around 100 ohms per square if not salicided. Because of their significant resistance dependence on voltage salicide diffusion resistors are rarely used intentionally, but they exist on every MOSFET as parasitic resistors. Non-salicided diffusion resistors are often used for ESD protection due to their high resistance and easy process compatibility with MOSFETs.

Polysilicon resistors are often used in integrated circuits for their nearly constant resistance over voltage and temperature. With different doping polysilicon resistors can have a a large range of sheet resistance. A salicide polysilicon resistor has a sheet resistance usually below 10 ohms per square, but a non-salicide polysilicon resistor with counter-doping has a sheet resistance over 1K ohms per square. The resistance variation over temperature of a non-salicide polysilicon resistor is 20 times smaller than the other thin film resistors on chip. Polysilicon resistors are often used for RC filtering, linear feedback control, DC biasing, gain and gain variation control over process and temperature.

6. Summary

This chapter discussed the devices in CMOS. Although MOSFETs are the most important devices in CMOS IC design, other devices, such as inductors, capacitors, varactors and resistors are also very critical. Properties of each device, especially those related to the CMOS process, have been discussed and their applications in RF circuits have been briefly reviewed.

Chapter 3

LINEAR TRANSCONDUCTORS IN CMOS

A transconductor is a circuit that converts an input voltage into an output current. Most circuits in CMOS are based on transconductors. The quality of the transconductors determines the performance of circuits. One of the most important specifications of a transconductor is its linearity. This chapter discusses linear transconductor designs in CMOS.

1. Differential pair

A differential pair is the most commonly used transconductor cell in CMOS. As an example an NMOS differential pair is shown in Figure 3.1. When the devices' channel is sufficiently long, the output current follows Equation 3.1, where μ_n is the NMOS mobility, C_{ox} is the gate capacitance per unit area, W and L are the channel width and length respectively, I_{dc} is the tail current and V_{in} is the differential input voltage.

$$I_{out} = \mu_n C_{ox} \frac{W}{L} \cdot V_{in} \cdot \sqrt{\frac{I_{dc}}{\mu_n C_{ox} \frac{W}{L}} - \frac{V_{in}^2}{4}} \tag{3.1}$$

Because the common source voltage changes with the input amplitude, the output current is not completely linear. The IIV3 is given by Equation 3.2.

$$IIV_3 = \sqrt{\frac{32}{3} \cdot \frac{I_{dc}}{\mu_n C_{ox} \frac{W}{L}}} \tag{3.2}$$

A simple differential pair is the most basic and easy-to-apply transconductor. However, its linearity is not sufficient to most circuits that can tolerate very small distortion. Many techniques have been developed to improve the linearity of the basic differential pair.

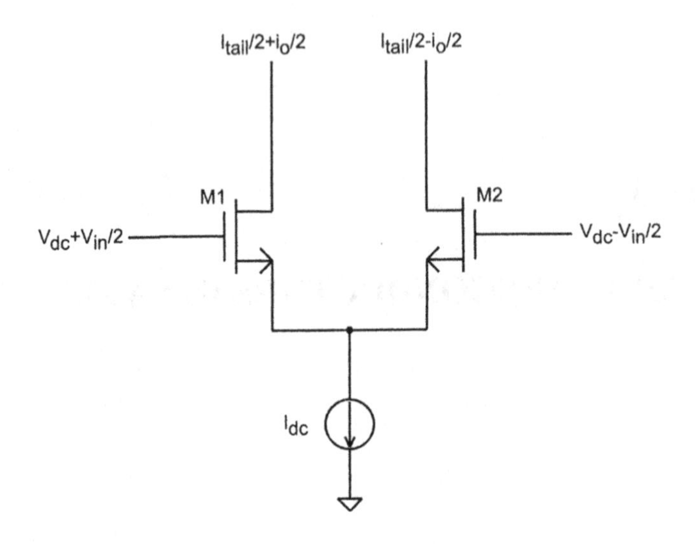

Figure 3.1. An NMOS differential pair

2. Bias offset cross-coupled differential pair

A transconductor with bias offset cross-coupled differential pairs is shown in Figure 3.2 [54]. Two differential pairs are biased with different DC voltages, and the outputs are cross-coupled to extract the linear signals. All the transistors of the cross coupled pairs (M1 to M4) have the same width (W) and length (L). The positive output current I_{o+} consists of the drain current from both M1 and M3, and the negative output current I_{o-} consists of the drain current from both M2 and M4, as are given by Equation 3.3 and Equation 3.4, respectively. M3 and M4 are cross-coupled with M1 and M2 with opposite inputs and shifted DC biases. The DC shift voltage V_B is realized by NMOS source followers M5 and M6. The output current is then given by Equation 3.5.

$$\begin{aligned} I_{o+} &= I_{d,M1} + I_{d,M3} \\ &= \frac{1}{2}\beta\left(V_{in+} - V_s - V_{tn}\right)^2 + \frac{1}{2}\beta\left(V_{in-} - V_B - V_s - V_{tn}\right)^2 \end{aligned} \quad (3.3)$$

$$\begin{aligned} I_{o-} &= I_{d,M2} + I_{d,M4} \\ &= \frac{1}{2}\beta\left(V_{in-} - V_s - V_{tn}\right)^2 + \frac{1}{2}\beta\left(V_{in+} - V_B - V_s - V_{tn}\right)^2 \end{aligned} \quad (3.4)$$

Where,

$$\beta = \mu_n C_{ox}\frac{W}{L}$$

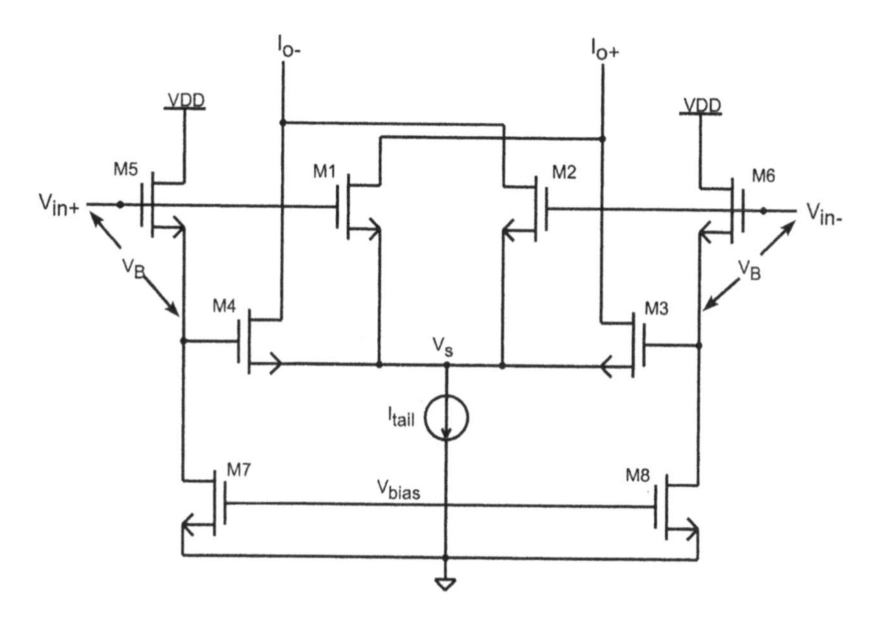

Figure 3.2. Cross-coupled differential pair

$$\begin{aligned} i_{out} &= I_{o+} - I_{o-} \\ &= \beta V_B V_{in} \end{aligned} \tag{3.5}$$

The linearity of the cross-coupled pairs is determined by the source followers. If a high impedance current source is provided to the source follower and $\frac{g_{m5}}{C_{gs4}}$ is much higher than the signal frequencies, V_B is considered to be constant. Therefore, a linear current output is achieved.

3. Source degenerated transconductor

Source degenerated resistors are often used to improve transconductors. Negative feedback is provided by the source degeneration resistors so that the effective transconductance is approximated by the conductance of the resistors, hence forming a linear transconductor.

3.1 Linear resistor degeneration

A differential pair with source degeneration resistors is shown in Figure 3.3. Usually polysilicon resistors are used for their superior linearity. The effective transconductance is given by Equation 3.6, where g_m is the transconductance

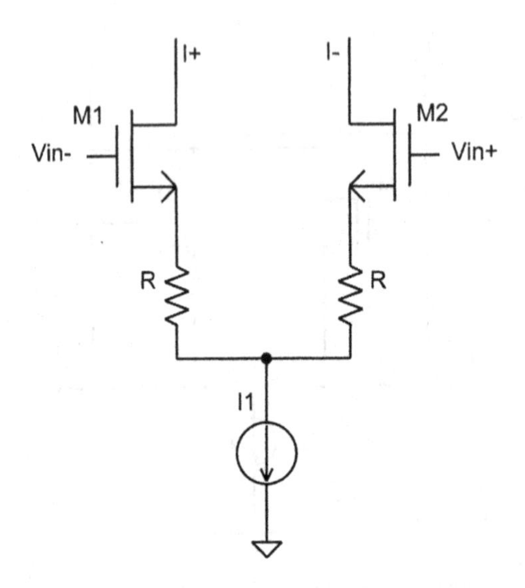

Figure 3.3. Source degenerated differential pair

of the NMOS and is given by Equation 3.7 when the channel is sufficiently long.

$$g_{meff} = \frac{g_m}{1 + g_m R} \tag{3.6}$$

$$g_m = \mu_n C_{ox} \frac{W}{L} V_{od} \tag{3.7}$$

Where, V_{od} is the over-drive voltage. When $g_m R \gg 1$, the effective transconductance is approximately equal to $\frac{1}{R}$ and the distortion is compressed by $g_m R$.

One drawback of the transconductor cell in Figure 3.3 is the head room reduction due to the DC voltage drop on the source degenerated resistors. To solve this problem two current sources can be used instead of one, as is shown in Figure 3.4. Only AC current flows through the source degenerated resistors; therefore, there is no DC voltage drops across the resistors, but AC negative feedback still exists and g_m distortion is compressed by $g_m R$.

As discussed above it is beneficial to use a large g_m to improve the linearity and keep the same effective transconductance. A "super MOS′′ is usually used to replace a single MOSFET to provide A times higher transconductance, where A is the gain of the extra amplifiers [19]. Figure 3.5 shows an example of a pair of "super NMOS′′ with a source degeneration resistor.

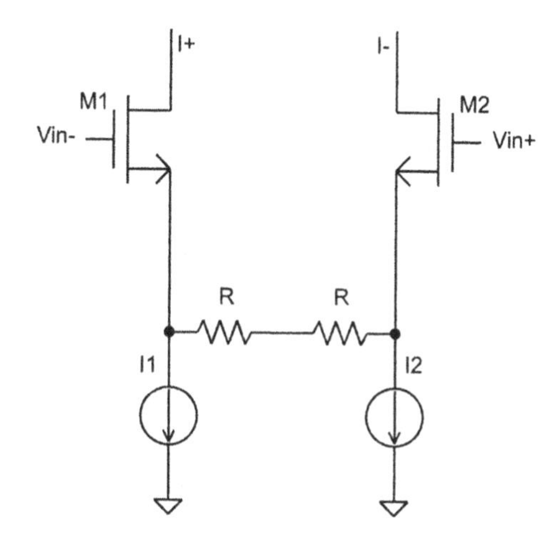

Figure 3.4. Source degenerated differential pair with two current sources

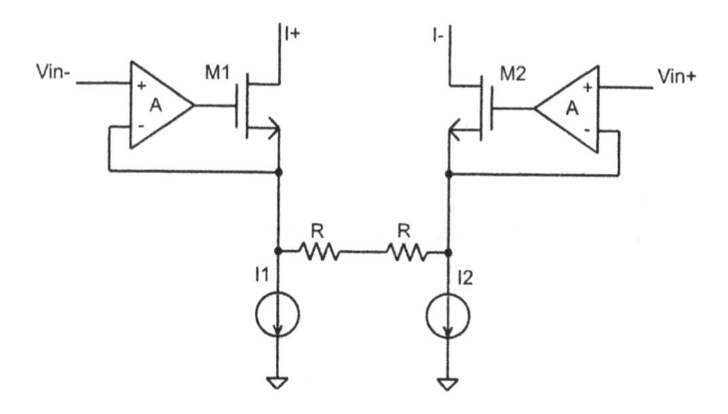

Figure 3.5. Source degenerated differential pair of "super MOS"

3.2 Triode-mode transistor degeneration

As discussed in Chapter 2 a MOSFET in triode mode has a linear response over drain-to-source voltage, and its *on* resistance is given by Equation 3.8. Some differential pairs use MOSFETs in triode mode instead of polysilicon resistors for source degeneration, as is shown in Figure 3.6 [24]. Because MOSFETs match each other better than MOSFETs match polysilicon resistors, a differential pair with triode transistor degeneration can provide a better linear transconductor than the polysilicon resistor degenerated differential pair.

$$R_{on}\left|_{V_{ds}\ll V_{od}} \simeq \frac{1}{\mu_n C_{ox}\frac{W}{L}\left(V_{gs}-V_{th}\right)}\right. \tag{3.8}$$

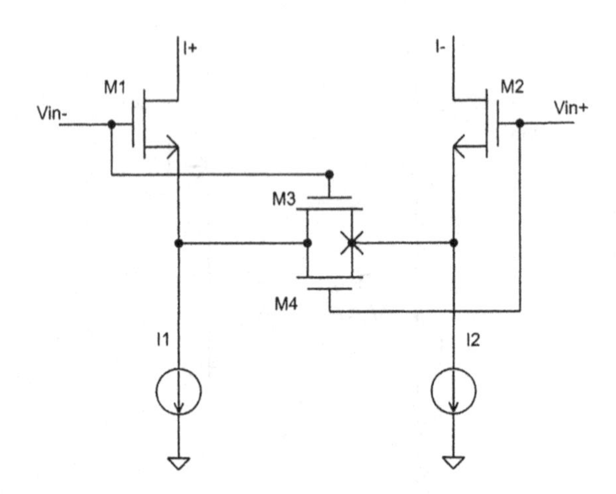

Figure 3.6. Differential pair degenerated by transistors in triode mode

3.3 Degeneration with constant V_{gs}

The linearity of a source degenerated differential pair in Figure 3.4 can be further improved if V_{gs} of the input transistors is kept constant. A constant V_{gs} transconductor cell has been proposed in bipolar [34], which is shown in Figure 3.7. With negative feedback to provide a constant drain current for the input transistors, V_{gs} of the MOS devices is forced to be constant over a large input range. Therefore, the input voltage is transferred across the linear resistor to produce a linear output current.

Both polysilicon resistors and triode mode transistors can be used to provide source degeneration in CMOS. As is discussed above polysilicon resistors have a better linearity, while triode mode transistors have the benefits of better matching and resistance tuning. Figure 3.8 shows an example with triode mode transistor degeneration [18].

4. Differential pair with a constant sum of V_{gs}

A differential pair can be linearized by keeping the sum of V_{gs} of the input transistors constant. The differential output current of a long channel CMOS pair is given by Equation 3.9.

$$\begin{aligned} I_{out} &= I_+ - I_- \\ &= \frac{1}{2}\beta\left(V_{gs+} + V_{gs-}\right)\left(V_{gs+} - V_{gs-}\right) \end{aligned}$$

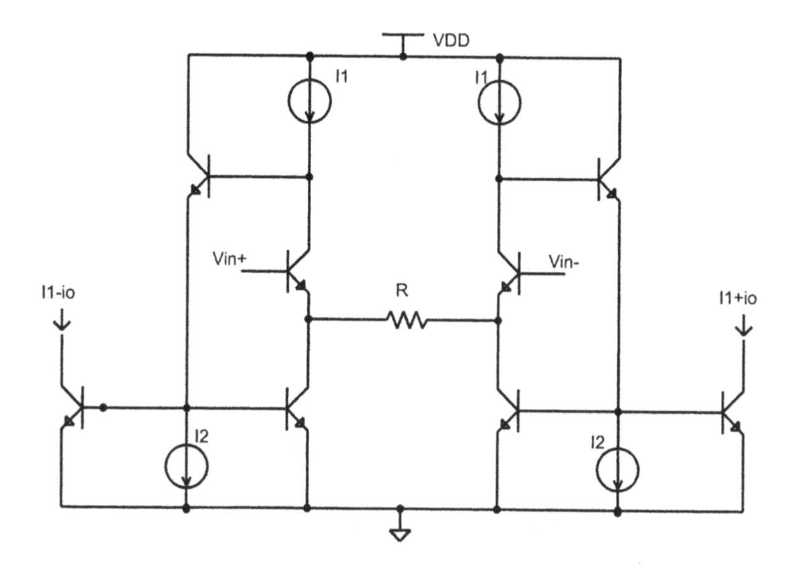

Figure 3.7. Source degenerated differential pair with constant V_{gs}

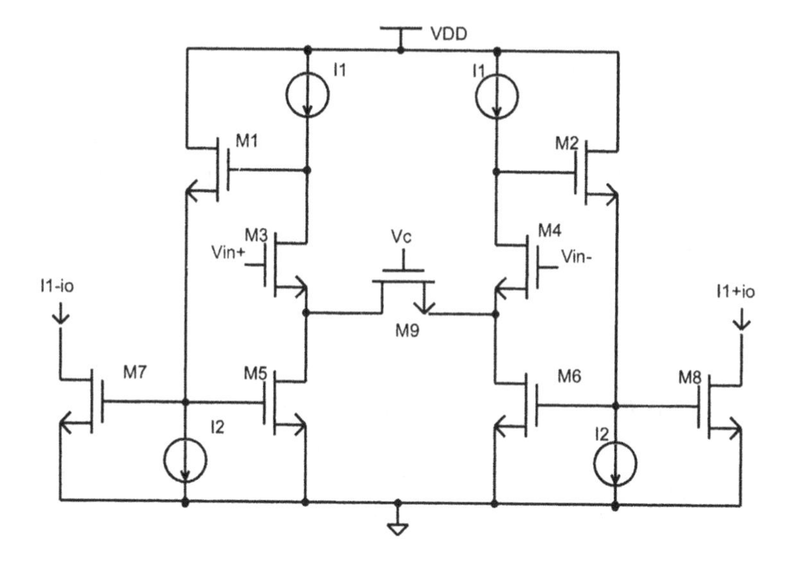

Figure 3.8. Transistor degenerated differential pair with constant V_{gs} in CMOS

$$= \frac{1}{2}\beta\left(V_{gs+} + V_{gs-}\right)V_{in} \tag{3.9}$$

The output consists of the product of the sum of V_{gs} and the difference of V_{gs}. The difference of V_{gs} equals the input voltage. If the sum of V_{gs} is constant, then

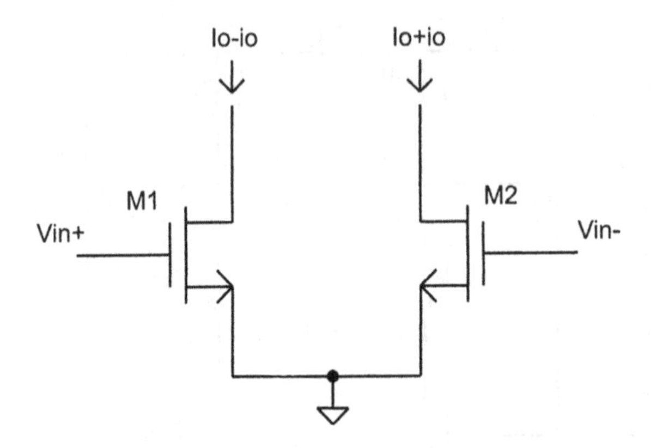

Figure 3.9. Source grounded differential pair

the output current is perfectly linear. Several techniques have been developed to design a linear differential pair transconductor with a constant sum of V_{gs} [50, 2, 49].

4.1 Source grounded pair

The simplest differential pair with constant sum of V_{gs} is a source grounded pair, as is shown in Figure 3.9 [50]. The sum of V_{gs} keeps constant as long as the input voltage is perfectly differential, i.e. there is no offset. One drawback of this design is a lack of input common mode rejection. In this aspect the source grounded pair is not a true differential pair, and is sometimes called a pseudo-differential pair.

4.2 Differential pair with floating voltage source

Figure 3.10 shows another design to keep the sum of V_{gs} constant [2]. A pair of floating DC voltage sources is used to shift the input voltage down. The sum of V_{gs} is given by Equation 3.10.

$$\begin{aligned} V_{gs+} + V_{gs-} &= [V_{in+} - (V_{in-} - V_{tn})] + [V_{in-} - (V_{in+} - V_{tn})] \\ &= 2V_{tn} \end{aligned} \tag{3.10}$$

The DC voltage sources can be realized by source followers, as is shown in Figure 3.11. With the floating voltage sources, a linear transconductor using constant sum of V_{gs} is achieved with good input common mode rejection.

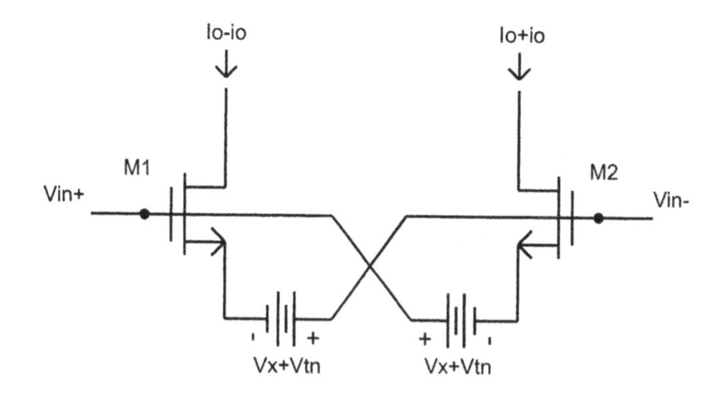

Figure 3.10. Differential pair with floating voltage sources

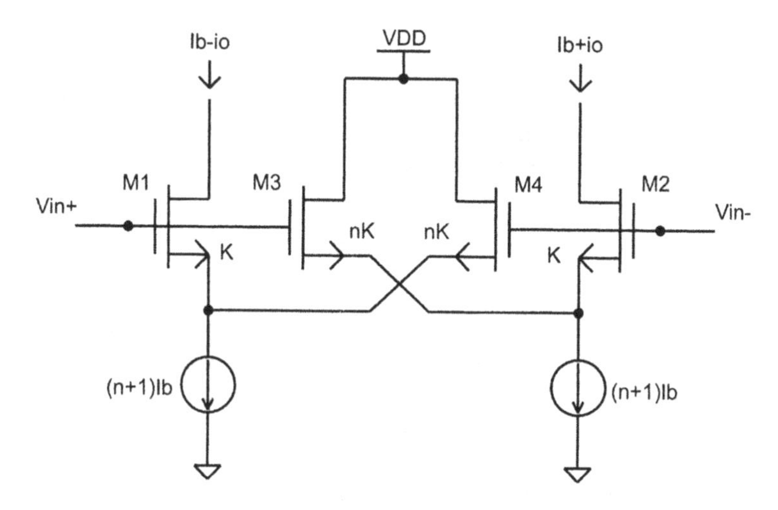

Figure 3.11. An example of differential pair with floating voltage sources

4.3 Differential pair with mobility compensation

The linear performance of the differential pair with constant sum of V_{gs} is based on the square-law I-V characteristic of the MOSFET. In reality this square-law behavior deviates from the ideal behavior due to channel velocity saturation and mobility degradation. The impact of channel velocity saturation can be reduced by increasing the channel length. However, mobility degradation is a function of the thickness of the gate oxide and becomes worse with faster processes. A more general I-V relation of a MOSFETis given in Equation 3.11 [49].

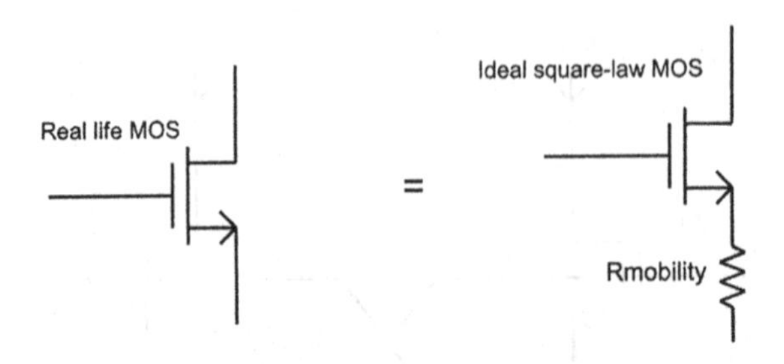

Figure 3.12. Effective model of NMOS with mobility degradation

$$I_d = \frac{1}{2}\beta \frac{V_{od}^2}{1 + \frac{1}{LE_{sat}} V_{od}} \tag{3.11}$$

Effectively the mobility degradation can be modeled as an ideal square-law MOSFET with a source degenerated resistor, as is shown in Figure 3.12 [16]. Because of the mobility degradation the transistors operate under a sub-square law. The degenerated resistor value is given by Equation 3.12.

$$R_{mobility} = \frac{1}{LE_{sat}\beta} \tag{3.12}$$

In order to compensate for the mobility degradation, positive feedback is used, as is shown in Figure 3.13. The loop gain of the positive needs to be carefully controlled so that it is sufficient to cancel the effective mobility feedback resistor but small enough to avoid instability.

5. Cross-coupled differential pairs with harmonic cancellation

This section presents a new transconductor with harmonic cancellation. When two differential pairs are cross-coupled with scaled inputs between each other, the signal harmonics can be cancelled at the output. Figure 3.14 shows the cross-coupled differential pairs with harmonic cancellation. The transconductance of the primary differential pair (M1 and M2) can be given by Equation 3.13 when the input is small and only 3rd order harmonic dominates the distortion, where g_m is given by Equation 3.7, and α is the 3rd order distortion factor. Likewise the transconductance of the auxiliary differential pair is given by Equation 3.14 with a certain scaling factor.

$$g_{m,pri}(V_{in}) = g_m \left(1 + \alpha V_{in}^2\right) \tag{3.13}$$

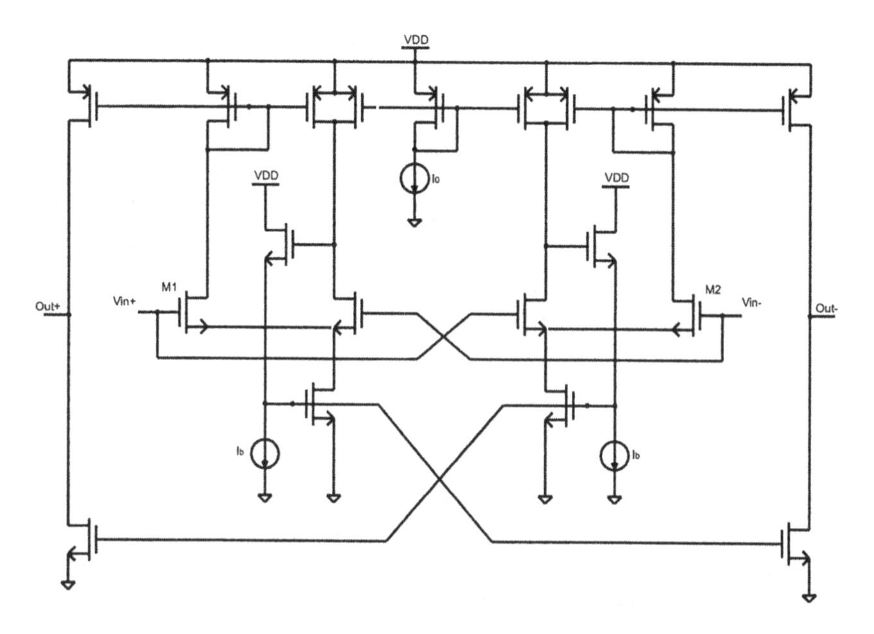

Figure 3.13. Differential pair with mobility compensation

$$g_{m,aux}(V_{in}) = \frac{1}{B^3} g_m \left(1 + \alpha V_{in}^2\right) \tag{3.14}$$

When the two differential pairs are cross-coupled and the auxiliary differential pair has a scaled input as that of the primary differential pair, the total transconductance is given by Equation 3.15.

$$\begin{aligned} g_{m,tot}(V_{in}) &= g_{m,pri}(V_{in}) - B \cdot g_{m,aux}(B \cdot V_{in}) \\ &= g_m \left(1 - \frac{1}{B^2}\right) \end{aligned} \tag{3.15}$$

Note that the 3rd order harmonic is cancelled at the output of the cross-coupled differential pairs, hence the effective total transconductance is linearized. This technique has benefits of high speed and general applicability to any process. The detailed discussion is covered in Chapter 4.

6. Summary

Several CMOS transconductors have been reviewed in this chapter. A differential pair is the most popular transconductor, but has limited linearity. Source degeneration may be added to a differential pair to improve its linearity. Either resistors or triode-mode transistors can be used for source degeneration, and

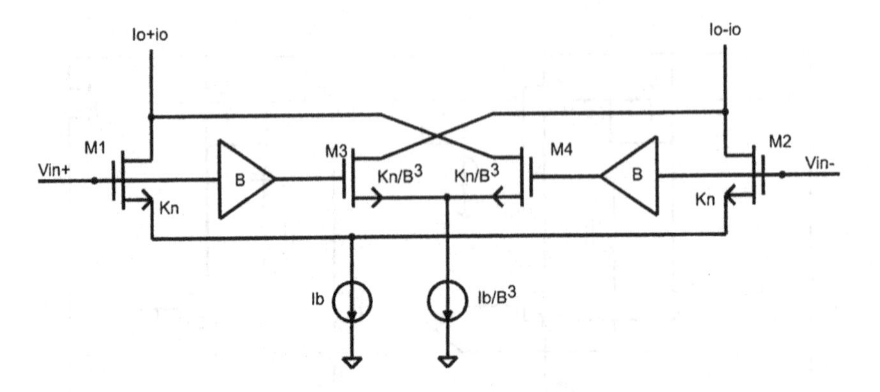

Figure 3.14. Differential pair with harmonic cancellation

a constant V_{gs} can further improve the linearity. The linearization technique of using a constant sum of V_{gs} has also been discussed. By using floating voltage sources or simply grounding the common source, the linearity of the differential pair can be greatly improved. The distortion due to the mobility degradation can also be compensated for by a positive feedback circuit. Most of the transconductors use feedback control or use long channel devices and are therefore not suitable for RF applications. The linearization technique using harmonic cancellation is most suited to RF circuits.

Chapter 4

LINEARIZATION WITH HARMONIC CANCELLATION

Modern wireless communication systems require circuits that operate at very high frequencies with high linearity. Although CMOS is the technology of choice for digital integration, it is seldom preferred for high performance linear RF circuits due to its limited ability to handle interferers. As discussed in Chapter 3 there are several linearization techniques used in CMOS design so that high linearity analog circuits can be integrated with digital circuits. However, those techniques are not suitable for very high frequency operation.

This chapter presents a linearization technique using harmonic cancellation. This technique can be applied to most circuits, such as LNAs and mixers, without any *a priori* knowledge of the nonlinear characteristic of the circuit and without sacrificing significant power, noise performance or gain.

1. Linearization in zero-memory weakly nonlinear systems

The proposed linearization technique is a feedforward technique in which the unwanted harmonics are cancelled. Although the harmonic cancellation technique can be generally applied to the weakly nonlinear systems with or without memory, it is more intuitive to be introduced through the zero-memory systems.

The transfer function of a weakly nonlinear system without memory (or feedback) can be expressed by Equation 4.1, where A is the gain factor, α_1, α_2,...,α_n are the power order coefficients, and C is the output offset[1].

$$y(x) = Ax\left(1 + \alpha_1 x + \alpha_2 x^2 + ... + \alpha_n x^n + ...\right) + C \qquad (4.1)$$

Here, the output is assumed to be a function of the input only. Although there may be other interferers at the output, they are uncorrelated to the input and can be considered as noise with respect to the input.

Note that all the signal ratios between the output harmonics, such as Ax, $\alpha_1 x$, $\alpha_2 x^2$, etc., depend on the input signal x; therefore, it is possible to isolate every harmonic including the completely linear component Ax. Theoretically, to characterize an arbitrary curve, an infinite number of values may be required. However, in most practical cases only a limited number of harmonics are needed to describe a system to a given accuracy. For example, to sufficiently describe the performance of a nonlinear system with up to the nth harmonic, n values are required to characterize every harmonic coefficient. As shown in Equation 4.2, all the power order coefficients, $A, \alpha_1, \alpha_2, ..., \alpha_{n-1}$, can be resolved, if n values of $y(x_1), y(x_2), ..., y(x_n)$ are provided.

$$\begin{aligned} y(x_1) &= Ax_1\left(1+\alpha_1 x_1+\alpha_2 x_1^2+...+\alpha_{n-1}x_1^{n-1}\right) \\ y(x_2) &= Ax_2\left(1+\alpha_1 x_2+\alpha_2 x_2^2+...+\alpha_{n-1}x_2^{n-1}\right) \\ &\vdots \\ y(x_n) &= Ax_n\left(1+\alpha_1 x_n+\alpha_2 x_n^2+...+\alpha_{n-1}x_n^{n-1}\right) \end{aligned} \tag{4.2}$$

Here, the output offset C is ignored as RF circuits are usually AC coupled. However, if the output offset C also needs to be considered, then one more value is required.

In particular, the linear gain A is given by Equation 4.3, where D is the determinant of the matrix $\{x_i^j\}_{nxn}$, i=1,...,n and j=1,...,n, and D_1 is the determinant of a similar matrix where the x_j's are replaced by $y(x_j)$, as shown in Equation 4.4 and Equation 4.5, respectively.

$$A = \frac{D_1}{D} \tag{4.3}$$

$$D = \begin{vmatrix} x_1 & x_1^2 & \cdots & x_1^n \\ x_2 & x_2^2 & \cdots & x_2^n \\ \cdots & \cdots & \cdots & \cdots \\ x_n & x_n^2 & \cdots & x_n^n \end{vmatrix} \tag{4.4}$$

$$D_1 = \begin{vmatrix} y(x_1) & x_1^2 & \cdots & x_1^n \\ y(x_2) & x_2^2 & \cdots & x_2^n \\ \cdots & \cdots & \cdots & \cdots \\ y(x_n) & x_n^2 & \cdots & x_n^n \end{vmatrix} \tag{4.5}$$

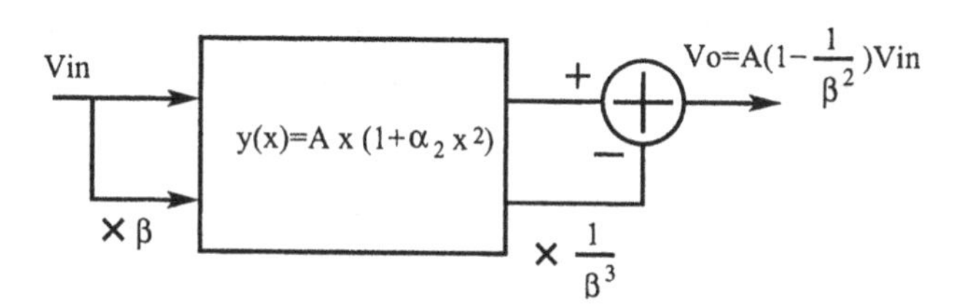

Figure 4.1. Linearization of the system with 3rd power order nonlinearity

More importantly Equation 4.3 indicates that the linear term Ax can be extracted from the nonlinear outputs.

Based on the above analysis a new technique has been developed to extract the linear component from the outputs. The extraction is realized by canceling all the other harmonics other than the fundamental by additional feedforward signal paths with different magnitude inputs.

1.1 3rd order harmonic cancellation

In most weakly nonlinear systems with differential design the 3rd harmonic is the most significant; therefore, a simplified scheme can be applied [52]. Such nonlinear circuits can be approximated by Equation 4.6.

$$y(x) = Ax(1 + \alpha_2 x^2) \tag{4.6}$$

If an auxiliary signal path with β times gain also passes through an identical system, then the 3rd harmonic of the primary output can be cancelled by subtracting $1/\beta^3$ of the auxiliary output. The procedure is shown in Figure 4.1. It can also be expressed as follows

$$\begin{aligned}
y_{primary}(x) &= Ax(1 + \alpha_2 x^2) \\
y_{auxiliary}(\beta x) &= A\beta x(1 + \alpha_2 \beta^2 x^2) \\
y(x) &= y_{primary}(x) - \frac{1}{\beta 3} y_{auxiliary}(\beta x) \\
&= A(1 - \frac{1}{\beta 2})x
\end{aligned} \tag{4.7}$$

Note that theoretically the 3rd harmonic caused by the 3rd power order term can be completely canceled, and hence the improvement in linearity can be infinite. However, in practice the nonlinearity cancellation is limited by device mismatch, higher power order terms ($\geq$5) and other out-of-cancellation nonlinearity.

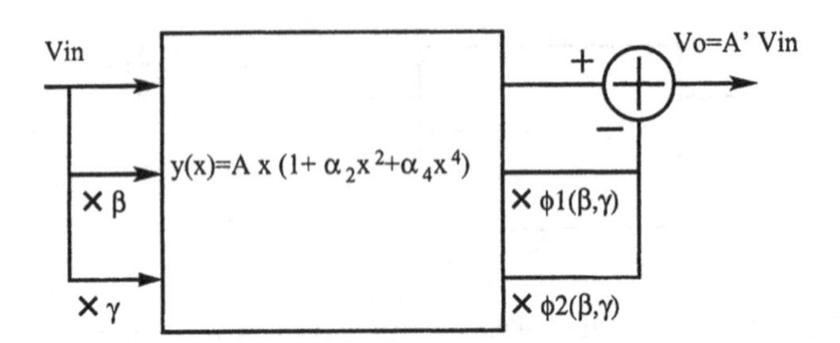

Figure 4.2. Linearization of the system with 3rd and 5th power order nonlinearity

1.2 Higher order harmonic cancellation

Most weakly nonlinear systems have the intermodulation outputs contributed from not only 3rd order distortion but also 5th or higher order distortion, such as the $A\alpha_4 x^5$ term in Equation 4.1. The time response of a nonlinear system with only the fundamental, the 3rd power order and the 5th power order, to a single tone, is given by Equation 4.8.

$$y(t) = Ax_0cos(\omega t)\left[1 + \alpha_2 x_0^2 cos^2(\omega t) + \alpha_4 x_0^4 cos^4(\omega t)\right] \tag{4.8}$$

$$\begin{aligned} y(t) \quad = \quad & Ax_0[(1 + \frac{3}{4}\alpha_2 x_0^2 + \frac{5}{8}\alpha_4 x_0^4)cos(\omega t) \\ & +(\frac{1}{4}\alpha_2 x_0^2 + \frac{5}{16}\alpha_4 x_0^4)cos(3\omega t) \\ & +\frac{1}{16}\alpha_4 x_0^4 cos(5\omega t)] \end{aligned} \tag{4.9}$$

There are the 3rd harmonics from both the 3rd power order and the 5th power order, as is given by Equation 4.9. At low input levels the 3rd harmonic contribution from the 5th power order is much less than that from the 3rd power order, which is the normal case in most RF receiver circuits. However, for some circuits with very high input levels, the 3rd harmonic contribution from the 5th power order is comparable to or even greater than that from the 3rd power order. For example, the input level for power amplifiers is typically more than -10 dBm. In such cases we have to consider canceling the 5th power order as well. In that case, another auxiliary feedforward signal path has to be created to provide further harmonic cancellation, as shown in Figure 4.2.

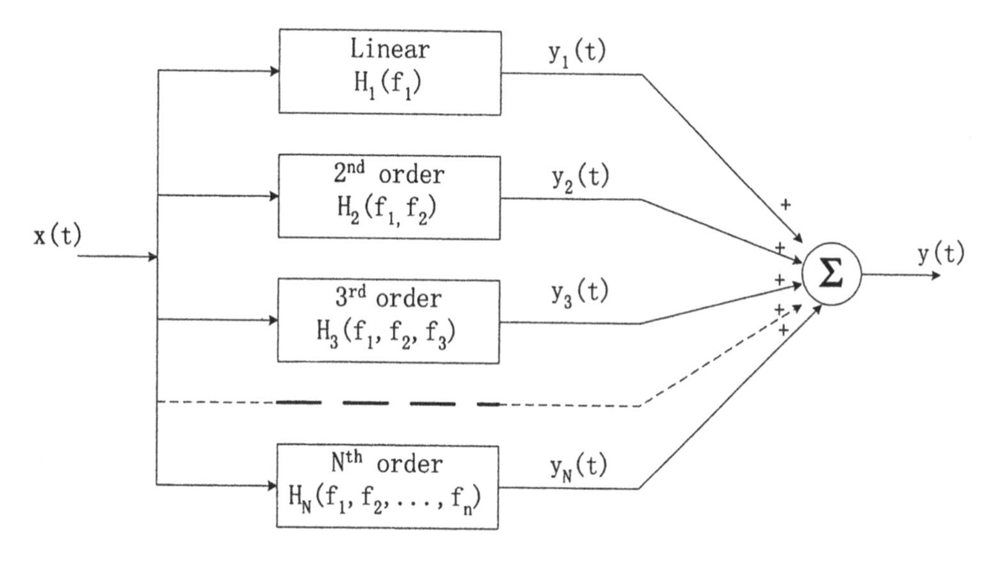

Figure 4.3. A weakly nonlinear system expressed in the form of a Volterra series

2. Linearization in weakly nonlinear systems with memory

The harmonic cancellation technique can be also applied to the weakly nonlinear systems with memory (or feedback), which are more general in real life applications.

A weakly nonlinear system can be expressed in the form of a Volterra series, as is shown in Figure 4.3 [23, 44, 36]. The output signal consists of the linear term, $y_1(t)$, and the sum of the higher order nonlinear terms, $y_2(t), \cdots, y_n(t)$,

$$y(t) = \sum_{n=1}^{N} y_n(t) \tag{4.10}$$

Any general input can be modeled as a sum of different sinusoidal signals (i.e, its Fourier components), as given by

$$x(t) = \frac{1}{2} \sum_{q=-Q}^{Q} E_q \, exp(j2\pi f_q t) \tag{4.11}$$

where E_q is the complex voltage or current of the qth tone, and f_q is the frequency of the qth sinusoidal signal. The nth order term of the nonlinear output is given by

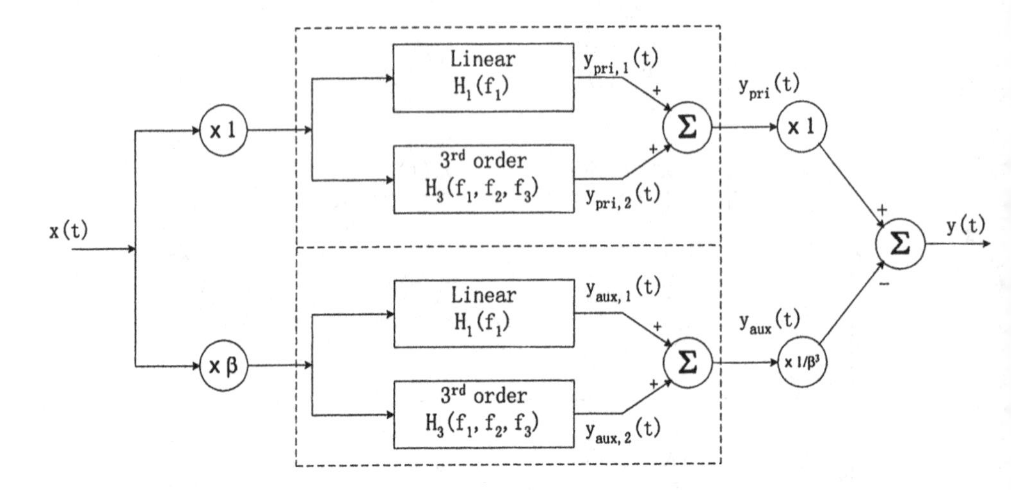

Figure 4.4. Block diagram of the linearization technique for 3rd order harmonic cancellation

$$
\begin{aligned}
y_n(t) &= \frac{1}{2^n} \sum_{q_1=-Q}^{Q} \cdots \sum_{q_n=-Q}^{Q} E_{q_1} \cdots E_{q_n} H_n(f_{q_1}, \cdots, f_{q_n}) \cdot \\
&\quad \exp\left[j2\pi(f_{q_1} + \cdots + f_{q_n})t\right] \qquad (4.12)
\end{aligned}
$$

If an input, $x(t)$, and its scaled value, $\beta\, x(t)$, go through similar weakly non-linear systems and their appropriately scaled outputs, 1 vs. β^{-n}, are subtracted from each other then the nth order nonlinear term is completely cancelled, as is shown below in Equation 4.13. The linearization procedure is best illustrated with the help of the block diagram shown in Figure 4.4.

$$
y_{new,n}(t) = y_{pri,n}\left(x(t)\right) - \frac{1}{\beta^n} \cdot y_{aux,n}\left(\beta x(t)\right) = 0 \qquad (4.13)
$$

It is important to note that although the nth order nonlinear term is cancelled, the other terms, such as the linear term, still exist. Therefore, with a number of similar procedures for each order, a completely linear term can be extracted, without any prior knowledge of the form of the nonlinear products. For example, for most fully differential slightly nonlinear circuits, the distortion is dominated by the 3rd order nonlinear products for small input levels.

For a two-tone input, the linear signal and the intermodulation terms at the output of the differential circuit can be expressed as shown in Equations 4.14 and 4.15, where θ_i (i=1 and 2) is the initial phase of the input signals, and ϕ_i is the phase delay of the linear transfer function $H_1(f)$ at the two input signal frequencies.

$$\begin{aligned} y_1(t) &= \left(1 - \frac{1}{\beta^2}\right) \{|E_1| \cdot |H_1(f_1)| \cos[2\pi f_1 t + \theta_1 + \phi_1] \\ &\quad + |E_2| \cdot |H_1(f_2)| \cos[2\pi f_2 t + \theta_2 + \phi_2]\} \end{aligned} \tag{4.14}$$

$$y_3(t) = y_{pri,3}(x(t)) - \frac{1}{\beta^3} y_{aux,3}(\beta x(t)) = 0 \tag{4.15}$$

Note that in the final output the 3rd harmonic has been completely cancelled. However, the linear output of the primary circuit is only reduced slightly in magnitude. Therefore, using this technique a completely linear transfer function is realized. Also note, that during this cancellation process, no knowledge of the nonlinear transfer function $H_3(f_{q1}, f_{q2}, f_{q3})$ is required.

In addition, the nonlinearity of the output load cannot be improved because it is outside the linearization system. In many cases highly linear passive loads are used to provide good linearity. However, more care has to be taken if active loads are used or if the passive loads do not provide sufficient linearity. Such nonlinearity can also be compressed if the output load is also included inside the system before subtraction or cancellation.

Though the linearization technique proposed here is very general and can be applied to most of the weakly nonlinear systems, for simplicity only 3rd order cancellation is discussed for the rest of the chapter.

3. Linearity and power consumption

In CMOS design, there is a tradeoff between linearity and power consumption. The transconductance linearity of short-channel MOSFETs has been studied and a theoretical curve of IIP3 versus power consumption has been reported [48]. As discussed in Chapter 2, the linearity of a MOSFET is defined by its input referred 3rd order intercept point (IIV3) as given by Equation 4.16, where V_{od} is the over-drive voltage, and α is given by Equation 4.17.

$$IIV_3 = \sqrt{\frac{8}{3}\frac{1}{\alpha}V_{od}\left(1 + \frac{1}{2}\alpha V_{od}\right)(1 + \alpha V_{od})^2} \tag{4.16}$$

$$\alpha = \theta + \frac{\mu_0}{2v_{sat}L} \tag{4.17}$$

Where, μ_0 is the mobility when the over-drive voltage is zero, L is the device channel length, v_{sat} is the channel carrier's saturation velocity, and θ is a modification factor of the mobility as shown in Equation 4.18.

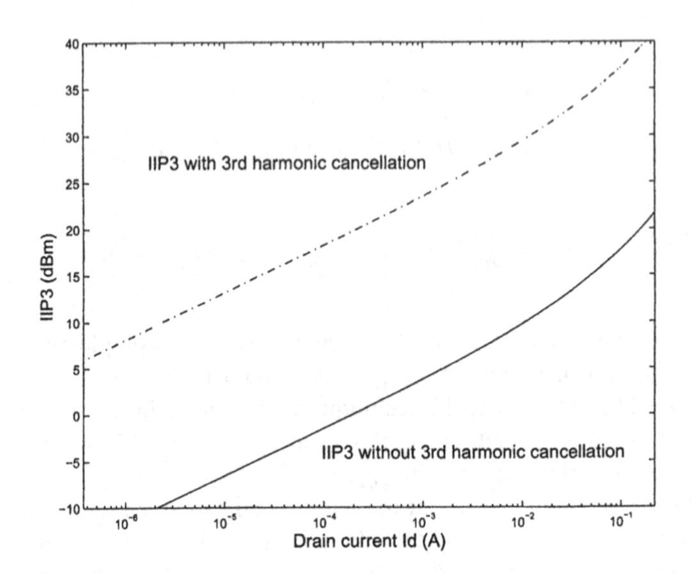

Figure 4.5. IIP3 comparison of MOSFET with and without 3rd harmonic cancellation

$$\mu_n = \frac{\mu_0}{1 + \theta V_{od}} \tag{4.18}$$

The DC current of a MOSFET is also a function of the over-drive voltage V_{od}, which is given by Equation 4.19 [49].

$$I_d = \frac{1}{2}\mu_n C_{ox}\frac{W}{L}\frac{V_{od}^2}{1 + \alpha V_{od}} \tag{4.19}$$

A relationship between the linearity and the DC power consumption is then defined by Equation 4.16 and Equation 4.19. A curve of IIP3 vs. power consumption is shown in Figure 4.5. Here a 50 Ω input impedance is assumed.

By applying the new harmonic cancellation technique to MOSFETs a new IIP3 vs. power consumption graph can be drawn, which is also plotted in Figure 4.5. Note that this technique not only reduces power consumption by several orders of magnitude for the same IIP3 specification, but also makes it feasible for MOSFET-based circuits including LNAs and mixers to meet high IIP3 requirements in a low voltage process.

4. Trade off between gain, power consumption and noise

Because additional auxiliary circuits have to be used to cancel the harmonics of the primary circuit at the output, there is additional power consumption and gain reduction. There is also additional noise contributed by the auxiliary circuits that needs to be considered.

4.1 Impact on power consumption

A power efficiency factor η is defined as the ratio of the power used by the primary circuits to the total power used by both the primary and the auxiliary circuits, as given by Equation 4.20.

$$\eta = \frac{P_{primary}}{P_{primary} + P_{auxiliary}} \tag{4.20}$$

In the case of 3rd power order cancellation only, as shown in Figure 4.1, and assuming the gain of the circuit is proportional to the size of the circuit, which is also proportional to the circuit's power consumption, the power efficiency factor is determined by the size of the auxiliary circuit, $\frac{1}{\beta^3}$, at the output, as given by Equation 4.21.

$$\eta = \frac{1}{1 + \frac{1}{\beta^3}} \tag{4.21}$$

If the auxiliary circuits are designed to be scaled versions of the primary circuit for better matching, and β is set to 2, then the power efficiency is 89%.

4.2 Impact on gain

Gain reduction caused by the auxiliary circuit can be defined as the ratio of the gain of the combined complete circuit to the gain of the primary circuit only, as is given by Equation 4.22.

$$\Gamma = \frac{A_{tot}}{A_{primary}} \tag{4.22}$$

In the case of 3rd power order cancellation only, as shown in Figure 4.1, the gain reduction is given by Equation 4.23.

$$\Gamma = 1 - \frac{1}{\beta^2} \tag{4.23}$$

Particularly, if the primary circuit and the auxiliary circuit are chosen that β equals 2, then the gain is reduced by $\frac{1}{4}$ or 2.5 dB.

4.3 Impact on noise

The noise from the auxiliary circuits can not be cancelled because it is uncorrelated with either the input or the noise from the primary circuit. The additional noise contribution is determined by the size of the auxiliary circuit, $\frac{1}{\beta^3}$, i.e., it increases the overall noise by $\frac{1}{\beta^3}$. In the case of $\beta = 2$ as above, the noise is increased by 0.2 dB.

Although the auxiliary circuits consume addition power, reduce the gain of the primary circuit and contribute additional noise, these impacts are small compared to the potential linearity improvement.

5. Harmonic cancellation with imperfections

Although the unwanted harmonics of the signals can be cancelled completely in theory, the cancellation is limited by imperfections in real life. The degree of the cancellation is usually determined by the mismatch of the input signals and mismatch of the circuits' gain.

5.1 Harmonic cancellation with input mismatch

Input mismatch between the primary circuit and the auxiliary circuit has an impact on the 3rd harmonic cancellation. For example, a weakly nonlinear system is described by Equation 4.24.

$$y(x) = Ax(1 + \alpha_2 x^2) \tag{4.24}$$

If the auxiliary circuit has an input with a phase mismatch of θ and a magnitude mismatch of ε compared to the β times scaled input to the primary circuit, its single tone expression can be given by Equation 4.25, where $V_{in,pri}$ and $V_{in,aux}$ are the primary input voltage and the auxiliary input voltage, respectively, and ω is the operating angular frequency.

$$V_{in,aux}cos(\omega t) = \beta V_{in,pri}(1 - \varepsilon)cos(\omega t + \theta) \tag{4.25}$$

In a two-tone test, where both the primary inputs and the auxiliary inputs have two-tone signals with a phase and magnitude mismatch, the harmonic cancellation results in Equation 4.26, where V_{out} is the output voltage before any filtering and $V_{out,IM3}$ is the intermodulation product of the two fundamental tones.

$$\begin{aligned}
V_{in,pri} &= x[cos(\omega_1 t) + cos(\omega_2 t)] \\
V_{in,aux} &= \beta x(1-\varepsilon)[cos(\omega_1 t + \theta) + cos(\omega_2 t + \theta)] \\
V_{out} &= Ax\{[cos(\omega_1 t) - \beta^{-2}(1-\varepsilon)cos(\omega_1 t + \theta)] \\
&\quad +[cos(\omega_2 t) - \beta^{-2}(1-\varepsilon)cos(\omega_2 t + \theta)]\} \\
&\quad +\alpha_2 x^3\{[cos(\omega_1 t) + cos(\omega_2 t)]^3 \\
&\quad -(1-\varepsilon)^3[cos(\omega_1 t + \theta) + cos(\omega_2 t + \theta)]^3\} \\
V_{out,IM3} &= \frac{3}{4}A\alpha_2 x^3\{cos[(2\omega_1 - \omega_2)t] - (1-\varepsilon)^3 cos[(2\omega_1 - \omega_2)t + \theta] \\
&\quad +cos[(2\omega_2 - \omega_1)t] - (1-\varepsilon)^3 cos[(2\omega_2 - \omega 1)t + \theta]\} \quad (4.26)
\end{aligned}$$

The 3rd harmonic cancellation factor ρ is defined as a ratio of the output intermodulation signal of the primary path, $V_{pri,IM3}$, to the output intermodulation signal after cancellation, $V_{out,IM3}$, as shown in Equation 4.27.

$$\begin{aligned}
\rho &= \frac{V_{pri,IM3}}{V_{out,IM3}} \\
&= \frac{1}{1-(1-\varepsilon)^3 cos\theta} \quad (4.27)
\end{aligned}$$

Figure 4.6 shows the intermodulation signal suppression versus different input magnitude and phase mismatch. For example, 27 dB suppression can be achieved with 0.1 dB magnitude mismatch and 9° phase mismatch.

5.2 Harmonic cancellation with circuit mismatch

In addition to input mismatch, gain mismatch between the primary circuit and the auxiliary circuit also affects the harmonic cancellation. Gain mismatch is usually caused by device parameter variation during the fabrication process, such as the threshold voltage V_{th}, the MOSFET channel length and width, and the metal or polysilicon resistance.

If the magnitude mismatch of the circuit gain is δ and its phase mismatch is ϕ, the intermodulation signal at the output is given by Equation 4.28, where $V_{out,pri,inband}$ and $V_{out,aux,inband}$ are the in-band output voltage of the primary circuit and the auxiliary circuit, respectively, and $V_{out,inband}$ is the in-band output voltage of the cross-coupled (primary and auxiliary) circuits. Other out-band signals are usually suppressed by the limited bandwidth at the output.

$$V_{in,pri} = x[cos(\omega_1 t) + cos(\omega_2 t)]$$

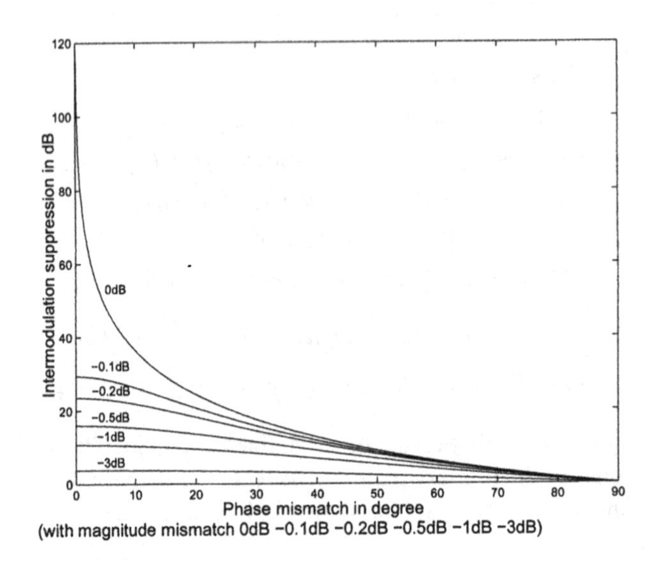

Figure 4.6. Impact of input mismatch on intermodulation suppression

$$
\begin{aligned}
V_{in,aux} &= \beta x[cos(\omega_1 t) + cos(\omega_2 t)] \\
V_{out,pri,inband} &= Ax(1 + \frac{9}{4}\alpha_2 x^2)\,[cos(\omega_1 t) + cos(\omega_2 t)] \\
&\quad + \frac{3}{4}\alpha_2 A x^3\,[cos(2\omega_1 t - \omega_2 t) + cos(2\omega_2 t - \omega_1 t)] \\
V_{out,aux,inband} &= \beta A(1-\delta)x(1 + \frac{9}{4}\alpha_2\beta^2 x^2)\,[cos(\omega_1 t + \phi) + cos(\omega_2 t + \phi)] \\
&\quad + \frac{3}{4}\alpha_2\beta^3 A(1-\delta)x^3[cos(2\omega_1 t - \omega_2 t + \phi) \\
&\quad + cos(2\omega_2 t - \omega_1 t + \phi)] \\
V_{out,inband} &\approx Ax[cos(\omega_1 t) - \beta^{-2}(1-\delta)cos(\omega_1 t + \phi) \\
&\quad + cos(\omega_2 t) - \beta^{-2}(1-\delta)cos(\omega_2 t + \phi)] \\
V_{out,IM3} &= \frac{3}{4}A\alpha_2 x^3\{cos[(2\omega_1 - \omega_2)t] \\
&\quad -(1-\delta)cos[(2\omega_1 - \omega_2)t + \phi]cos[(2\omega_2 - \omega_1)t] \\
&\quad -(1-\delta)^3 cos[(2\omega_2 - \omega 1)t + \phi]\}
\end{aligned} \tag{4.28}
$$

The harmonic cancellation with the circuit mismatch is given by Equation 4.29.

$$
\rho = \frac{1}{1 - (1-\delta)cos\phi} \tag{4.29}
$$

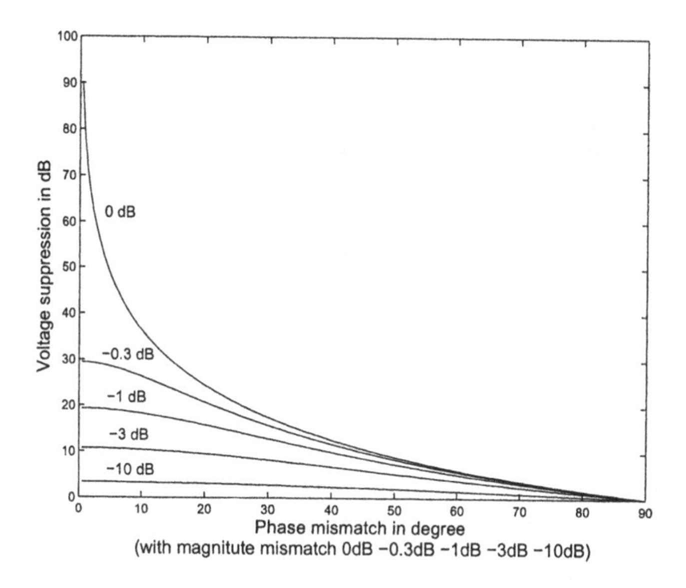

Figure 4.7. Impact of circuit mismatch on intermodulation suppression

Figure 4.7 shows the intermodulation signal suppression versus different circuit gain and phase mismatch. For example, 27 dB suppression can be achieved with 0.3 dB magnitude mismatch and 9° phase mismatch.

It has been well known that the device matching can be improved by proper layout, such as the central tapping layout and the interdigitated layout; and it is also improved by the area of the devices [33]. For example, the variance of parameter ΔP of a group of four cross-coupled transistors can be given by Equation 4.30. Here A_p and S_p are the area proportionality constant and the spacing proportionality constant for parameter P, respectively; D_x and D_y are the spacing between the devices in the diagonal directions of x and y; and WL gives the area of the devices.

$$\sigma^2(\Delta P) = \frac{A_p^2}{2WL} + S_p^2 \frac{D_x^2 D_y^2}{wafer\ diameter^2} \tag{4.30}$$

It is possible to obtain better than 1% device matching in current CMOS process. Therefore, the output linearity improvement can be up to 40 dB. However, as is discussed in the chapter, the linearity improvement by the harmonic cancellation in real life applications is always limited by the higher order (larger than 3) nonlinearity, the other nonlinear load at the output, which is outside the cancellation system, and the parasitic mismatch.

6. Summary

This chapter presented a linearization technique with harmonic cancellation. Because it doesn't require any *a priori* knowledge of the system, it can be applied to the weakly nonlinear circuits. Also because it is a feedforward cancellation, it has little impact on the speed of the circuits. Therefore, it is a preferred linearization technique for RF circuits. Although the distortion can be completely cancelled in theory, the cancellation is limited by the process mismatch. The impact of input mismatch and gain mismatch has also been discussed. With 1% process mismatch, the output distortion can be suppressed by up to 40 dB. Chapter 5 and 6 discuss applications of this technique to RF low noise amplifiers and mixers, respectively.

Notes

1 The gain factor A and the power order coefficients, α_1, α_2,...,α_n, contain the phase information if all the variables and the coefficients are expressed in complex format.

Chapter 5

LNA DESIGN IN CMOS

The low noise amplifier (LNA) is one of the key blocks in most high performance radio receivers. It is usually the first block of the receiver after the antenna. The LNA amplifies the input signals so that the noise generated by following blocks has little impact on the system signal-to-noise (SNR) ratio. If the gain of LNA is sufficient, the receiver's SNR is dominated by that of the LNA. A LNA has to have a low noise figure (NF).

Many modern radio receivers require not only a large SNR but also a large dynamic range, which requires that all the receiver blocks, including the LNA maintain good SNR for both small signals and large signals. The circuit's linearity is one of the limiting factors to its dynamic range, and it is one of the major specifications of LNA.

Common-gate amplifiers and common-source amplifiers are the most commonly used low noise amplifiers in radio frequencies. This chapter discusses these two types of LNAs, especially their gain, noise and linearity. The harmonic cancellation technique is applied to LNA designs, and the linearity improvement of this technique and other impacts are discussed.

1. Common-gate low noise amplifier

The core device in a common-gate LNA in CMOS is the common-gate MOSFET at the input. Because the LNA directly interfaces with the antenna, 50 Ω impedance matching is usually required at its input. A common-gate MOSFET can provide a low input impedance and is easy to match to 50 Ω over a very wide band, even without other passive components such as inductors. This advantage makes the common-gate LNA a preferable choice for many low cost and/or multi-band radios.

A typical common-gate LNA is shown in Figure 5.1. Its performance can be characterized by its gain, noise and linearity.

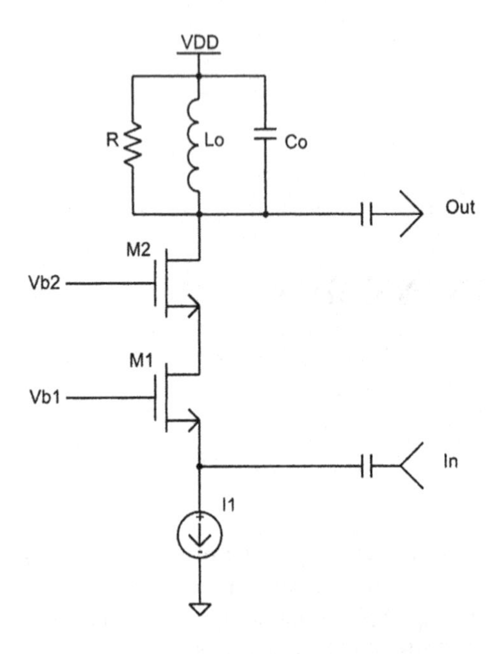

Figure 5.1. Common-gate LNA

1.1 Gain

The voltage gain of the common-gate LNA is given by Equation 5.1.

$$A_v = g_m Z_{out} \tag{5.1}$$

The gain is a function of transconductance g_m and the output impedance Z_{out}. As shown in Figure 5.1, the output impedance of the LNA is dominated by the parallel LC tank. Since the transconductance is nearly constant over frequency and the tank impedance peaks at its resonant frequency f_o, the LNA has its maximum gain at f_o. In most cases the inductor of the tank has a lower quality factor (Q) than the capacitor. For example, the Q of a typical on-chip inductor in a typical CMOS process is between 4 and 10, and the Q of a typical MIM (metal-insulator-metal) cap in CMOS is over 40. Therefore, the tank impedance is mostly limited by the quality of the inductor. The tank impedance at the resonant frequency is given by Equation 5.2, where ω_o is the resonant angular frequency, which equals $\sqrt{\frac{1}{L_o C_o}}$, Q_{L_o} is the quality factor of the tank inductor and L_o is the inductance of the tank inductor.

$$Z_{out} = Q_{L_o}\omega_o L_o \tag{5.2}$$

The voltage gain at the resonant frequency is then given by Equation 5.3.

$$A_v = g_m Q_{L_o}\omega_o L_o \tag{5.3}$$

The available power gain of an amplifier is a function of its voltage gain and both the input and the output impedance, as given by Equation 5.4.

$$A_p = A_v^2 \frac{Z_{in}}{Z_{out}} \tag{5.4}$$

The available power gain can only be achieved when the input impedance is matched to the source impedance. The input impedance of a common-gate LNA is given by Equation 5.5, where C_{gs} is the gate-to-source capacitance and ω_T is the MOSFET cut-off frequency.

$$\begin{aligned} Z_{in} &= \frac{1}{g_m + j\omega_o C_{gs}} \\ &= \frac{1}{g_m}\left(\frac{1}{1 + j\frac{\omega_o}{\omega_T}}\right) \end{aligned} \tag{5.5}$$

The input impedance has a bandwidth of ω_T. For most CMOS processes applicable to radio frequency operation the cut-off frequency is on the order of tens of giga Hertz. For example, the maximum cut-off frequency for the TSMC 0.25 μm CMOS process is higher than 45 GHz. Since most CMOS radios operate under 10 GHz, the input impedance of the common-gate LNA is dominated by the inverse of its transconductance and is constant over a wide range of frequencies. When the input transistor satisfies the condition in Equation 5.6, where Z_s is the source impedance at the input, which is 50 Ω for most RF circuits. The input impedance matches over a wide band .

$$g_m = \frac{1}{Z_s} \tag{5.6}$$

When both the input and the output are matched, the power gain of the common-gate LNA is given by Equation 5.7.

$$A_p = g_m \omega_o L_o Q_{L_o} \tag{5.7}$$

1.2 Noise

As discussed in Chapter 2, there are three major noise contributions in a MOSFET: channel thermal noise, flicker noise and induced gate noise [5]. Because the gate of the input MOSFET is connected to ground, and the LNA usually operates at much higher frequencies than the flicker noise corner frequency, the thermal noise dominates the total noise contribution in a common-gate LNA. The channel thermal noise is given by Equation 5.8 [17], where k is the Boltzmann's constant (1.38 x 10^{-23} J/K), T is the temperature in Kelvin, g_{d_0} is the zero-biased drain conductance, and γ is a bias-dependent factor.

$$i_{d,noise}^2 = 4kT\gamma g_{d_0} \tag{5.8}$$

When the MOSFET is in saturation, g_{d_0} is approximately equal to g_m. For a long channel MOSFET, γ is between $\frac{2}{3}$ and 1; when the MOSFET is in saturation, γ equals $\frac{2}{3}$, and when the MOSFET is in triode mode, γ equals 1. For a short channel MOSFET, γ is much higher than $\frac{2}{3}$ even in saturation. Depending on the bias, γ could be as high as two or three [47].

A noise factor (F) is defined as the input signal-to-noise ratio divided by the output signal-to-noise ratio, as shown in Equation 5.9. Noise figure (NF) is the noise factor expressed in dB. Equation 5.10 gives the noise factor for a common-gate LNA, where α is defined as the ratio of g_m over g_{d_0}.

$$F = \frac{SNR_{in}}{SNR_{out}} \tag{5.9}$$

$$F_{LNA} = 1 + \frac{\gamma}{\alpha} \tag{5.10}$$

When the input MOSFET is a long channel device in saturation, the NF of the LNA equals 2.2 dB.

The input matching of a LNA affects its noise figure. Any loss caused by input mismatch will add to the noise figure directly, as is given by Equation 5.11, where $Loss_{input}$ is the input loss caused by mismatch. Equation 5.12 gives $Loss_{input}$ from the input return loss S11. A bad input match can seriously degrade the noise figure.

$$NF_{LNA} = 10log(1 + \frac{\gamma}{\alpha}) + Loss_{input}(dB) \tag{5.11}$$

$$Loss_{input} = 1 - S11 \tag{5.12}$$

Because the input impedance of the common-gate LNA is a function of the device cut-off frequency, and the cut-off frequency is approximately inversely proportional to the square of the channel length, a shorter channel is preferred in the input transistor. The relation of the cut-off frequency and the channel length is given by Equation 5.13 [17], where μ_{eff} is the effective channel mobility, V_{od} is the over-drive voltage, and L is the channel length.

$$\omega_T \approx \frac{3}{2}\frac{\mu_{eff} V_{od}}{L^2} \tag{5.13}$$

Although the MOSFET's cut-off frequency decreases with the channel length, because the γ increases when the channel length decreases, the noise created by the input common-gate MOSFET increases as well [47]. Usually the input transistor is optimized to balance input matching and γ to provide the minimum noise figure.

1.3 Linearity

For those LNAs with linear load at the output the linearity is primarily determined by the input transconductors. The transconductance of a MOSFET in saturation is a function of the channel width (W) and length (L) and the over-drive voltage (V_{od}), as is given by Equation 5.14.

$$g_m = \mu_n C_{ox} \frac{W}{L} V_{od} \frac{1 + \frac{\alpha}{2} V_{od}}{(1 + \alpha V_{od})^2} \tag{5.14}$$

Equation 6.26 gives the definition of α, where μ_0 is the mobility when the over-drive voltage is zero, and θ is a process constant.

$$\alpha = \theta + \frac{\mu_0}{2 v_{sat} L} \tag{5.15}$$

The over-drive voltage of a common-gate MOSFET is also a function of the input voltage (V_{in}) and the threshold voltage (V_{th}), as shown in Equation 5.16, where V_b is the bias voltage of the gate.

$$V_{od} = V_b - V_{in} - V_{th} \tag{5.16}$$

The body effect causes the threshold voltage to change with the input voltage as well, as is given by Equation 5.17, where V_{th0} is the threshold voltage when the source-to-substrate voltage (V_{sb}) equals zero, γ_{th} is the body-effect constant and ϕ_f is the Fermi potential at the channel. This introduces more distortion into LNA.

$$V_{th} = V_{th0} + \gamma_{th}\left(\sqrt{V_{sb} + 2\phi_f} - \sqrt{2\phi_f}\right) \tag{5.17}$$

Connecting source and substrate together can reduce the distortion but increases the parasitic capacitance at the source. Without the body effect an IIV3 (input referred 3rd order intercept point in voltage) of a common-gate LNA is given by Equation 5.18 [48].

$$IIV_3 = \sqrt{\frac{8}{3}\frac{1}{\alpha}V_{od}\left(1 + \frac{1}{2}\alpha V_{od}\right)(1 + \alpha V_{od})^2} \tag{5.18}$$

In general, increasing the over-drive voltage and using a longer channel device can improve the linearity of a MOSFET; however, it also increases the power consumption and reduces the cut-off frequency.

1.4 Common-gate LNA with harmonic cancellation

The linearization technique of harmonic cancellation can improve the linearity of a common-gate LNA significantly with a slight trade-off of decreased gain and increased noise. Figure 5.2 shows an example of a common-gate LNA with harmonic cancellation.

The input to the auxiliary LNA is β times larger than that to the primary LNA. Since the auxiliary circuit only has to provide $\frac{1}{\beta^3}$ times gain of the primary circuit, its input device can be $\frac{1}{\beta^3}$ times smaller than that of the primary circuit if both the primary circuit and the auxiliary circuit have the same bias voltage. As a result, the input impedance of the auxiliary circuit is β^3 times higher than that of the primary circuit, which is shown in Equation 5.19, where $Z_{in,aux}$ and $g_{m,aux}$ are the input impedance and the transconductance of the auxiliary circuit, respectively, and $Z_{in,pri}$ and $g_{m,pri}$ are the input impedance and the transconductance of the primary circuit, respectively.

$$Z_{in,aux} = \frac{1}{g_{m,aux}} = \frac{\beta^3}{g_{m,pri}} = \beta^3 Z_{in,prim} \tag{5.19}$$

As discussed in Chapter 4, the third harmonics of the outputs of the primary circuit and the auxiliary circuit cancel to each other; therefore, the signal distortion at the cross-coupled output is suppressed by tens of dB. For example, assume there is 1% input mismatch caused by the process variation and another

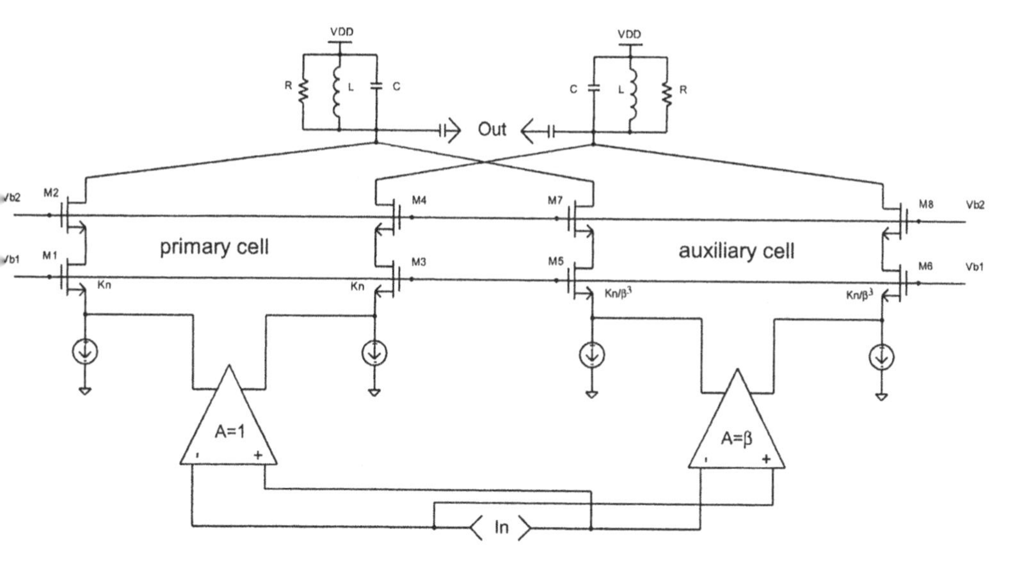

Figure 5.2. Common-gate LNA with harmonic cancellation

1% gain mismatch between the primary circuit and the auxiliary circuit, the total distortion at the output of LNA is suppressed by 26.4 dB. Compared to the gain degradation of 2.5 dB and the power consumption increase of slightly more than 40%, the improvement in linearity is significant. There is also a trade-off in the noise figure since more circuitry is added. However, because the size of the auxiliary circuit and the front amplifier are both much smaller than the primary circuit, the additional noise is much smaller than the noise generated by the primary circuit. For example, if β is 2, the noise figure at the output of LNA is increased by no more than than 0.6 dB.

2. Common-source low noise amplifier

A simple common-source amplifier is shown in Figure 5.3. A pseudo differential pair is used as the input transconductor cell. The input impedance is set by the gate-to-source capacitor so that additional tuning components are usually used to match it to the source impedance. However, for a common-source LNA the condition where the input impedance is matched to the source impedance is not necessarily the same condition where the noise figure is minimized [49]. A separate tuning circuit is required to minimize the noise figure. To provide freedom to both tune the input impedance and minimize the noise figure, two pairs of inductors are used, as is shown in Figure 5.4. The source-degeneration inductors connect the source of MOSFET and ground so as to provide an effective resistive input load without contributing additional noise. Gate inductors are used before the gate to optimize the noise figure.

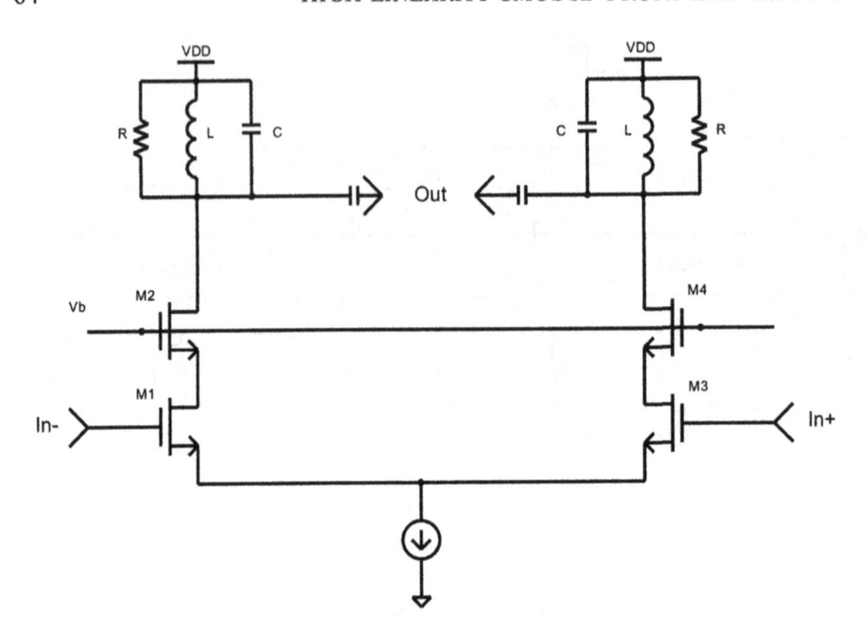

Figure 5.3. Simple common-source amplifier

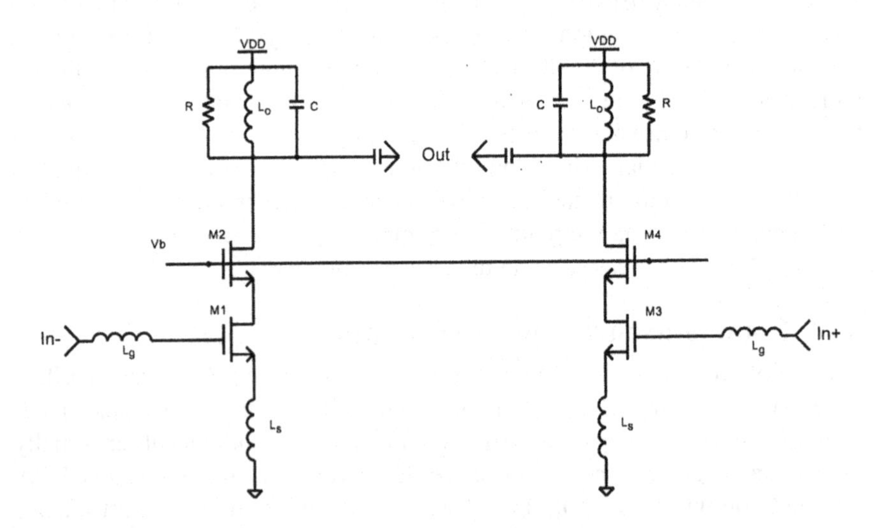

Figure 5.4. Inductive source-degenerated LNA

2.1 Gain

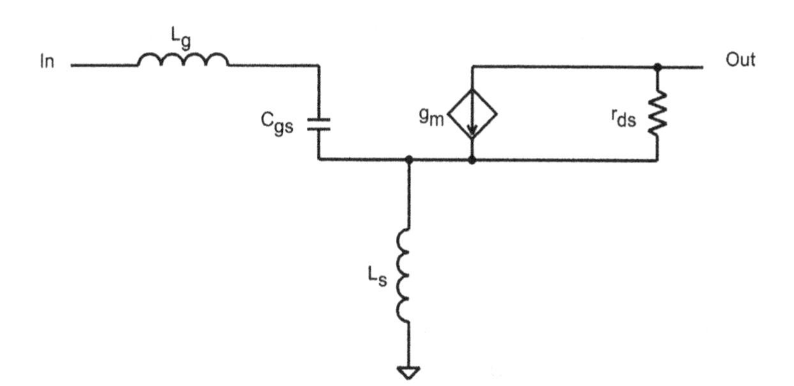

Figure 5.5. Small-signal model of an inductive source-degenerated LNA

A small-signal model of an inductive source-degenerated LNA is shown in Figure 5.5. Viewed from the input, the source-degeneration inductor L_s works effectively as a noiseless resistor, whose value is given by Equation 5.20. The effective transconductance is given in Equation 5.21, where ω_T is the cut-off angular frequency and equals $\frac{g_m}{C_{gs}}$.

$$R_{L_s} = \omega_T L_s \tag{5.20}$$

$$g_{m,eff} = Q_{in} g_m \tag{5.21}$$

Where, Q_{in} is the quality factor of the series LC tank at the input, which consists of a gate inductor L_g, a source-degenerated inductor L_s, a gate-to-source capacitor C_{gs} and an effective resistor R_{L_s}. The value of Q_{in} at the resonant frequency is given by Equation 5.22.

$$\begin{aligned} Q_{in} &= \frac{1}{\omega_o C_{gs} R_{L_s}} \\ &= \frac{1}{\omega_o C_{gs} \omega_T L_s} \end{aligned} \tag{5.22}$$

When the input impedance is matched to the source impedance R_s (usually 50 Ω), Q_{in} can be expressed as done in Equation 5.23.

$$Q_{in} = \frac{1}{\omega_o C_{gs} R_s} \tag{5.23}$$

From Equation 5.21 and Equation 5.23, the effective transconductance of an inductive source-degenerated LNA is given by Equation 5.24.

$$g_{m,eff} = \frac{\omega_T}{\omega_o}\frac{1}{R_s} \tag{5.24}$$

As shown in Equation 5.1 and Equation 5.4 the power gain is a function of the input transconductance, the input impedance and the output impedance. The output impedance is dominated by the parallel LC tank as given by Equation 5.2. When both the input and the output are impedance matched, the power gain A_p of an inductive source-degenerated LNA is given by Equation 5.25 .

$$A_p = \left(\frac{\omega_T}{\omega_o}\right)^2 \frac{Q_{L_o}\omega_o L_o}{R_s} \tag{5.25}$$

2.2 Noise

There are three major noise contributions in an inductive source-degenerated LNA, the gate thermal noise caused by the gate contact resistance and the gate inductor parasitice resistance, the channel thermal noise and the induced gate noise. Equation 5.26 gives an expression for the noise factor [17], where R_{L_g} is the parasitic resistance of L_g, R_g is the gate parasitic resistance and χ is the modification factor of the induced gate noise. The factor χ is given in Equation 5.27, where c is the correlation factor of the induced gate noise to the drain thermal noise (approximately j0.395 for long channel MOSFETs), and δ is the coefficient of the gate noise. The factor δ equals $\frac{4}{3}$ for a long channel device and could be as large as 2 for a short channel device.

$$F = 1 + \frac{R_{L_g}}{R_s} + \frac{R_g}{R_s} + \gamma\chi g_{d0} R_s \left(\frac{\omega_o}{\omega_T}\right)^2 \tag{5.26}$$

$$\chi = 1 + 2 \mid c \mid Q_{in}\sqrt{\frac{\delta\alpha^2}{5\gamma}} + \frac{\delta\alpha^2}{5\gamma}\left(1 + Q_{in}^2\right) \tag{5.27}$$

Derek Shaeffer and Thomas Lee proved in their paper [17] that there is an optimal value of the input series LC tank to give a minimum noise figure. Especially for a power constrained design, Q_{in} is optimized to a constant value, as is given by Equation 5.28.

$$Q_{in,opt} = |c| \sqrt{\frac{5\gamma}{\delta}} \left[1 + \sqrt{1 + \frac{3}{|c|^2} \left(1 + \frac{\delta}{5\gamma} \right)} \right] \approx 3.9 \quad (5.28)$$

Equation 5.23 shows the monotonic relation between Q_{in} and C_{gs} for a matched input. Therefore, an optimal size of the input device can be calculated from $Q_{in,opt}$, as given by Equation 5.29, where C_{ox} is the gate capacitance per unit area and L is the device length.

$$W_{opt} = \frac{1}{Q_{in,opt}} \frac{1}{\omega_o C_{ox} L R_s} \quad (5.29)$$

2.3 Linearity

Similar to a common-gate LNA, the linearity of the inductive source-degenerated LNA is also dominated by its input transconductor g_m. However, because the voltage across gate and source is Q_{in} times higher than the input voltage before the gate inductor, the effective IIV3 is Q_{in} times smaller than that of a common-gate LNA, as is given by Equation 5.30.

$$IIV_3 = \frac{1}{Q_{in}} \sqrt{\frac{8}{3} \frac{1}{\alpha} V_{od} \left(1 + \frac{1}{2} \alpha V_{od} \right) (1 + \alpha V_{od})^2} \quad (5.30)$$

When the input impedance is matched to R_s, IIP3 is then given by Equation 5.31.

$$IIP_3 = \frac{1}{Q_{in}^2 R_s} \frac{8}{3} \frac{1}{\alpha} V_{od} \left(1 + \frac{1}{2} \alpha V_{od} \right) (1 + \alpha V_{od})^2 \quad (5.31)$$

Compared to the common-gate LNA, the inductive source-degenerated LNA trades off its linearity for its superior noise performance.

2.4 Common-source LNA with harmonic cancellation

The linearization technique in Chapter 4 can also be applied to the common-source LNA. As shown in Figure 5.6, simple linear circuits are used to generate signals with a different amplitude to the auxiliary circuit than the primary circuit, i.e., the voltage across gate and source of the auxiliary circuit is β times that of the primary circuit. Because the gain of the auxiliary circuit is $\frac{1}{\beta^3}$ times smaller

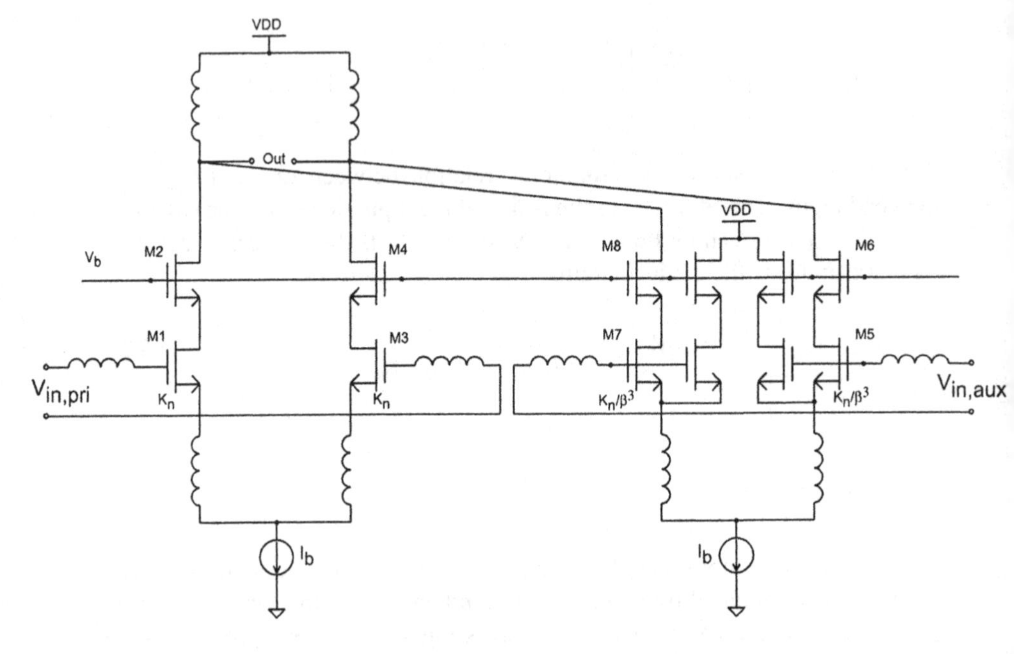

Figure 5.6. Source-degenerated LNA with harmonic cancellation

than that of the primary circuit, at the output the third harmonic and the major intermodulation signals are suppressed significantly, while the linear portion of the signals is only slightly impacted. Because the size of the auxiliary circuit is much smaller than that of the primary circuit, the additional noise contribution is very limited as well.

2.5 An example of a common-source LNA

A common-source LNA with a source-degenerated inductor is designed with the harmonic cancellation technique [51]. A simplified circuit diagram is shown in Figure 5.6. An auxiliary circuit is used to cancel the 3rd harmonic.

The LNA is designed to operate at 900 MHz and have 2 dB noise figure, 18 dB gain and +5 dBm IIP3. Using a β of 2, the auxiliary LNA requires one-eighth the gain of the primary LNA for proper cancellation of the 3rd harmonic. In order to achieve better matching, the same input transistor size and the same drain current has been used for both the primary LNA and the auxiliary LNA, but only one-eighth of the auxiliary LNA's output current is used to subtract from the primary LNA's output current. Therefore, the power consumption is doubled. With 1% matching the improvement of the output linearity is less

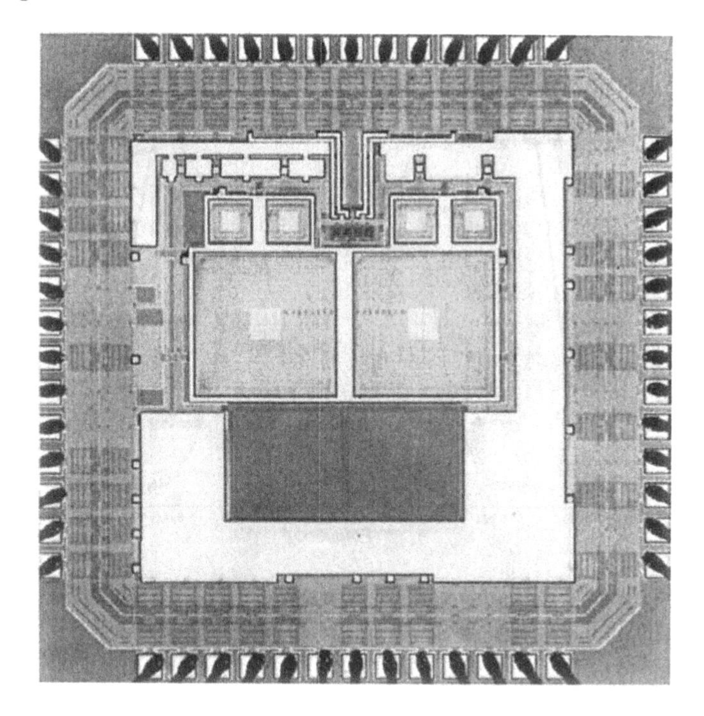

Figure 5.7. Die photograph of a source-degenerated LNA with harmonic cancellation

than 40 dB due to the nonlinear component of the transistor r_{ds} that is not suppressed by this technique. Compared to the similar LNA design but without the harmonic cancellation scheme, the simulated results of this LNA shows a 14 dB improvement of IIP3. Since the noise contribution due to the auxiliary LNA is limited to one-eighth of the original LNA design, the noise figure is expected to increase by only 0.2 dB.

A prototype of a high linearity LNA has been fabricated in the TSMC 0.35 μm process available through MOSIS. The die micro-photograph is shown in Figure 5.7. The external gate inductors and the external matching networks have been used to match the input and the output.

The two-tone technique was used to measure IIP3. A comparison of the output spectral responses with and without 3rd harmonic cancellation is shown in Figure 5.8 for a relatively large input of -22 dBm (due to the limited dynamic range of the spectrum analyzer). Figure 5.9 gives the spectral comparison of the intermodulation signal for a smaller input of -32 dBm. The intermodulation product is suppressed by 27 dB for this input level.

The measured IIP3 is plotted in Figure 5.10. The measured IIP3 without 3rd harmonic cancellation is +5 dBm, and the measured IIP3 with 3rd harmonic cancellation is +18 dBm. Note that the 3rd harmonic has been greatly sup-

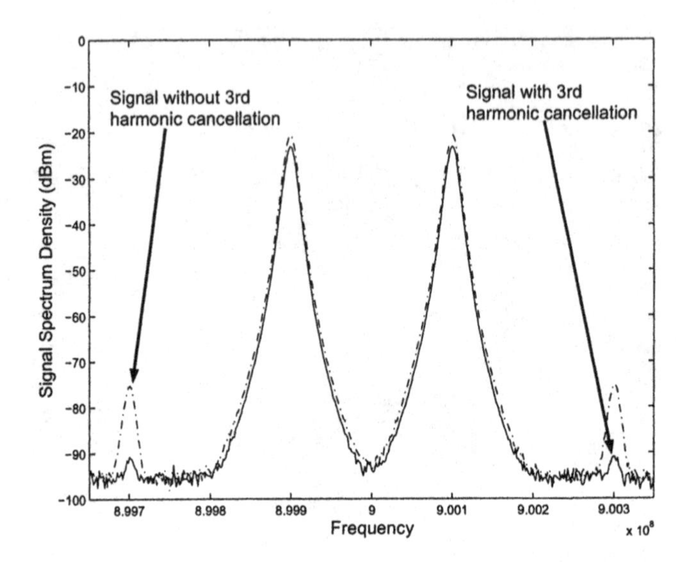

Figure 5.8. Two tone test with and without 3rd harmonic cancellation

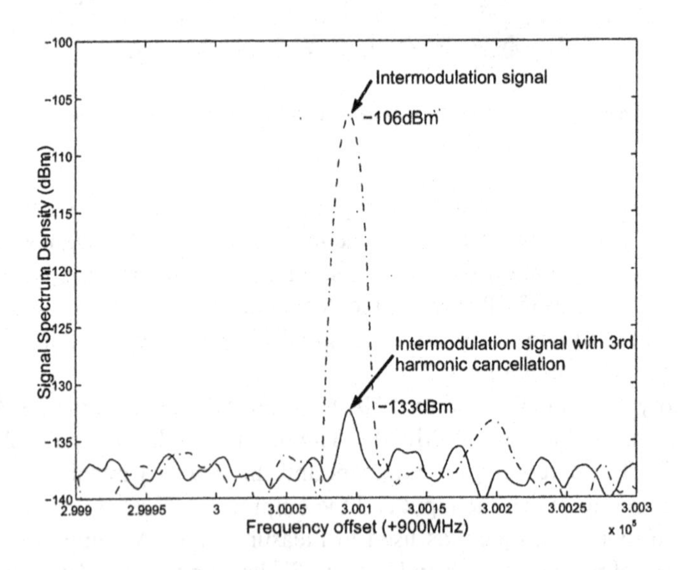

Figure 5.9. Comparison of the intermodulation signals for smaller inputs

pressed and that for high input levels the slope of the intermodulation product increases, which suggests a higher power order contribution to the nonlinearity. For all inputs lower than -25 dBm our LNA maintains a very low distortion level, which makes this LNA attractive for most high performance wireless communication systems.

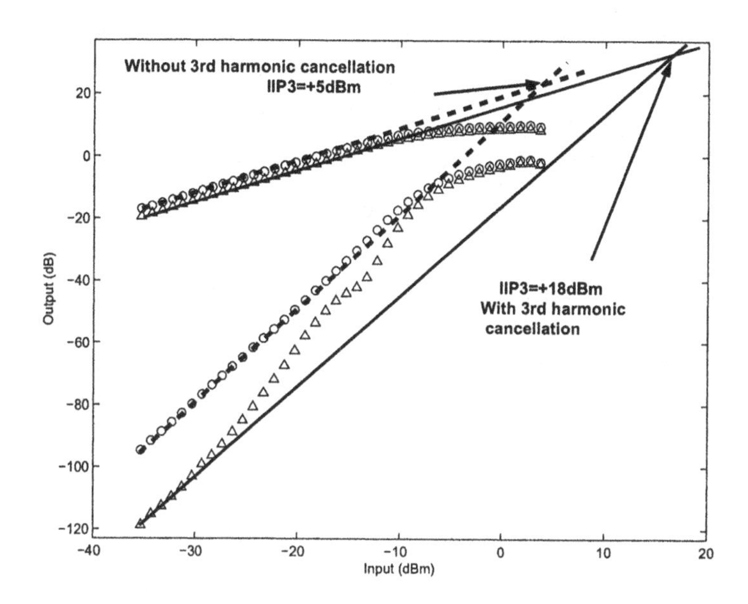

Figure 5.10. Comparison of IIP3 with and without 3rd harmonic cancellation

One of the most commonly used thumb rules in the differential circuit design is that the P1dB of the circuit is about 9.6 dB lower than its IP3. However, that rule is only valid when the 3rd order nonlinearity dominates. One interesting result from the 3rd order harmonic cancellation is that the circuit's P1dB is no longer 9.6 dB lower than its IP3. As is shown in Figure 5.10, although the circuit's IP3 is increased by 13 dB, its P1dB is almost not changed. That makes the difference of the P1dB and the IP3 is close to 22 dB.

A comparison of the realized IIP3 vs. power consumption of this LNA with previously published LNA designs in CMOS [17, 38–1, 14, 31, 6, 3, 35], bipolar [32, 39, 7, 30] and GaAs [25, 40] is shown in Figure 5.11. The linearity of this LNA is not only higher than any other reported CMOS LNA, but also superior to SiGe LNAs and GaAs LNAs. The measurement results are summarized in Table 5.1.

3. Summary

This chapter discussed low noise amplifiers in CMOS. The common-gate LNA provides a wide band input match and the inductive source-degenerated LNA has a lower noise figure. With help of a series gate inductor and a source-degenerated inductor the LNA can minimize its noise figure and match its input impedance to the source impedance independently. However, compared to the common-gate LNA the input of the inductive source-degenerated LNA can only be matched in a narrow band. Although the inductive source-degenerated LNA

Performance Parameters	LNA	LNA with 3rd Harmonic Cancellation
Operating Frequency	900 MHz	
Supply Current (mA)	7.5	15
Noise Figure (dB)	2.6	2.8
IIP3 (dBm)	+5	+18
Gain (dB)	18	15

Table 5.1. Summary of measured results

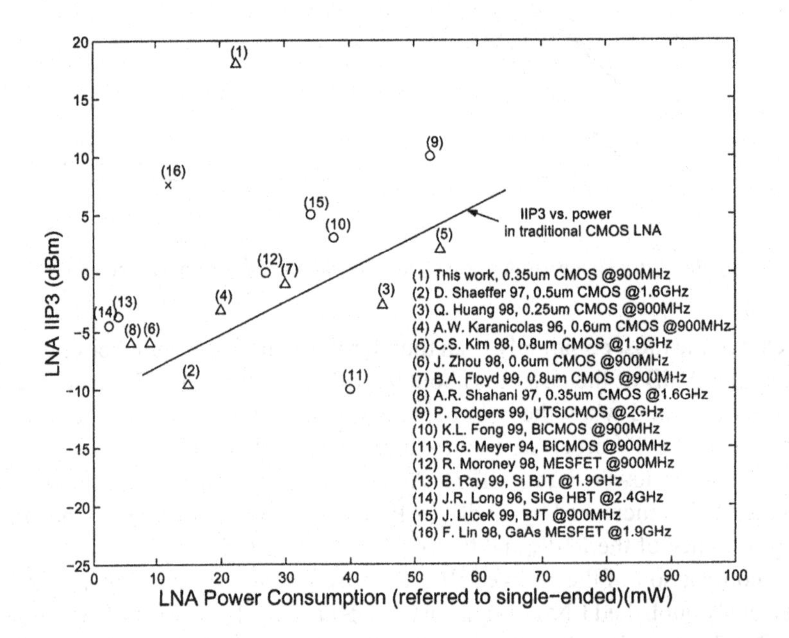

Figure 5.11. Comparison of reported IIP3 vs. power consumption

has a lower noise figure, it has a lower IIP3 as well. The linearization technique of harmonic cancellation has been applied to an inductive source degenerated LNA. The measurement results show an improvement of IIP3 by 13 dB. The linearized LNA has 15 dB gain, 2.8 dB NF and +18 dBm IIP3.

Chapter 6

DOWN-CONVERSION MIXER DESIGN IN CMOS

In a radio transceiver the bandwidth of the useful information is usually in the order of a few kilo-Hertz or mega-Hertz. However, a radio usually transmits signals through the air at much higher frequencies. For example, a transceiver for 802.11a wireless LAN sends and receives signals at 5 to 6 GHz with only 20 MHz bandwidth. A transmitter needs up-conversion mixers to transmit signals using higher frequency carriers; and a receiver needs down-conversion mixers to extract the information from giga-Hertz signals. The mixer is one of the key blocks in a transceiver design and its dynamic range limits the entire receiver's or transmitter's dynamic range. In this chapter only down-conversion mixers are discussed.

A down-conversion mixer often follows a low noise amplifier (LNA). Due to the gain of the LNA, the mixer can have a higher noise figure but also requires a higher IIP3. The noise figure of the mixer is usually designed to be smaller than the gain of LNA so that its noise contribution is smaller than that of the LNA. Because the input signal to the mixer is higher than that to the LNA, the linearity requirement of the mixer is higher than that of the LNA. In most of cases the mixer's IIP3 limits the IIP3 of the front-end.

This chapter discusses passive mixers and balanced Gilbert mixers, especially regarding their conversion gain, noise and linearity. The harmonic cancellation technique is also applied to a Gilbert mixer, which follows a LNA. This technique improves the dynamic range of the receiver front-end (LNA and down-conversion mixer) by more than 10 dB, but with a small increase of power consumption.

1. Passive mixer

Passive mixers are the most common and simplest down-conversion circuits. They were also the first to be used in receivers at the early stage of radio design.

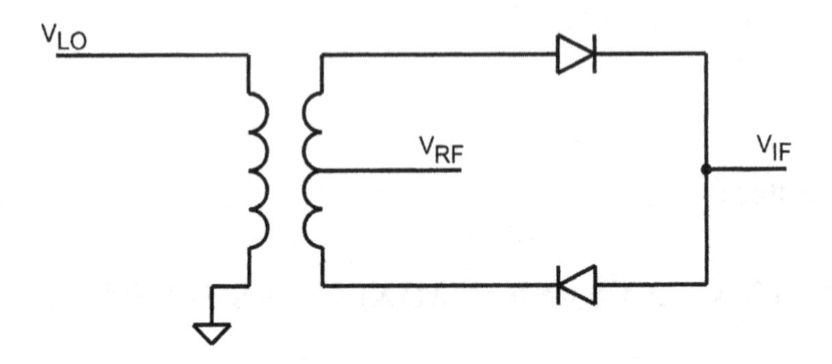

Figure 6.1. Diode mixer

A transformer and two diodes make one of the simplest passive mixers, as is shown in Figure 6.1. This mixer has good isolation between LO and IF, and also between RF and LO. But RF signals goes directly to the IF output.

Because a switch can be realized by a simple MOSFET, a passive voltage commutating mixer can be easily implemented in CMOS. Figure 6.2 shows a double-balanced passive mixer. When the LO signals are large enough, only the two transistors in the diagonal positions are on during every half period and the other two transistors are off. The RF signals are switched between the IF nodes through different pairs of switches in the diagonal positions in the LO period.This mixer can be considered as a voltage multiplier of the RF signals and an unit-amplitude square-wave of +1 and -1 at the LO frequency.

Besides the IF frequency mixing product there are many other harmonics at the IF output. Because it is a balanced design, all of the even order harmonics are cancelled. Additional filters are often applied at the IF output to reduce the unwanted harmonics.

1.1 Conversion gain

The conversion gain of a down-conversion mixer is defined as the signal power or voltage at the IF output divided by the signal power or voltage at the RF input, as given by Equation 6.1.

$$\begin{aligned} A_p &= \frac{P_{out,IF}}{P_{in,RF}} \\ A_v &= \frac{V_{out,IF}}{V_{in,RF}} \end{aligned} \tag{6.1}$$

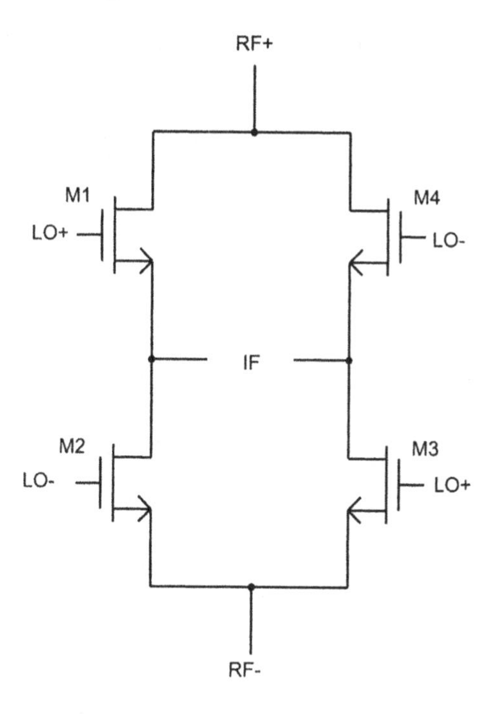

Figure 6.2. Double-balanced passive mixer

A. Shahani has shown in his paper [3] that the output of this passive down conversion mixer can be given by Equation 6.2, where $g_T(t)$ is the time-varying Thevenin-equivalent conductance viewed from IF, g_{Tmax} and $\overline{g_T}$ are the maximum value and the average value of $g_T(t)$, respectively, and m(t) is the mixing function. Equation 6.3 gives the function m(t), where T_{LO} is the period of LO signals.

$$v_{IF}(t) = v_{RF}(t) \cdot \left[\frac{g_T(t)}{g_{Tmax}} \cdot m(t)\right] \cdot \frac{g_{Tmax}}{\overline{g_T}} \tag{6.2}$$

$$m(t) = \frac{g(t) - g(t - T_{LO}/2)}{g(t) + g(t - T_{LO}/2)} \tag{6.3}$$

When the LO is a square wave, the maximum voltage conversion gain is $\frac{2}{\pi}$, i.e., -3.9 dB; when the LO is a sinusoidal wave, the maximum voltage conversion gain is $\frac{\pi}{4}$, i.e., -2.1 dB. However, in practice it is difficult to reach the maximum conversion gain. Furthermore, because the output impedance at IF is usually much higher than the input impedance at RF in IC design, the actual power conversion gain is usually below -10 dB.

A large LO drive is required for this type of mixer so that the passive transistors can be switched on and off periodically. Although there is no DC power used by the passive mixer itself, the LO drive circuits consume a large amount of power to provide sufficient LO swing at the input of the mixer switches. Therefore, the total power consumption has to include the power used by the LO drive circuits.

1.2 Noise

A low noise amplifier is usually used before a down-conversion mixer; therefore, the required noise figure of the mixer is much higher than that of the LNA. As shown in Equation 6.4, the noise figure of the LNA adds to the noise figure of the system directly, but the contribution of the noise figure of the mixer is compressed by the gain of LNA.

$$NF_{total} = 1 + (NF_{LNA} - 1) + \frac{NF_{mixer} - 1}{A_{LNA}} + ... \quad (6.4)$$

As a passive device, its noise figure nearly equals its power loss. The single side band (SSB) noise figure of the passive mixer is usually higher than 10 dB.

1.3 Linearity

Linearity is one of the major concerns of a down-conversion mixer. Both the signals and the interferers are amplified by LNA before they pass through the mixer. Many interferers are too close to the signals to be filtered out by on chip RF filters and those interferers can be much stronger than the desired signal. Therefore, the down-conversion mixer requires much better linearity than the LNA does. As shown in Equation 6.5 the distortion contributed by the mixer is A_{LNA} time higher than that of LNA.

$$\frac{1}{IIP_{3,total}} = \frac{1}{IIP_{3,LNA}} + \frac{A_{LNA}}{IIP_{3,mixer}} + ... \quad (6.5)$$

If the switches in a passive mixer are ideal switches, there is no distortion created by the mixer. However, because the resistance of the switches depends on not only the LO drive voltage but also the input voltage, the signal is distorted by the switches. For example, both the instantaneous conductance of the switch ($g_T(t)$) and the mixing function ($m(t)$) are the function of the input ($v_{RF}(t)$) so that at the output there is not only a linear term of $v_{RF}(t)$ but also higher order

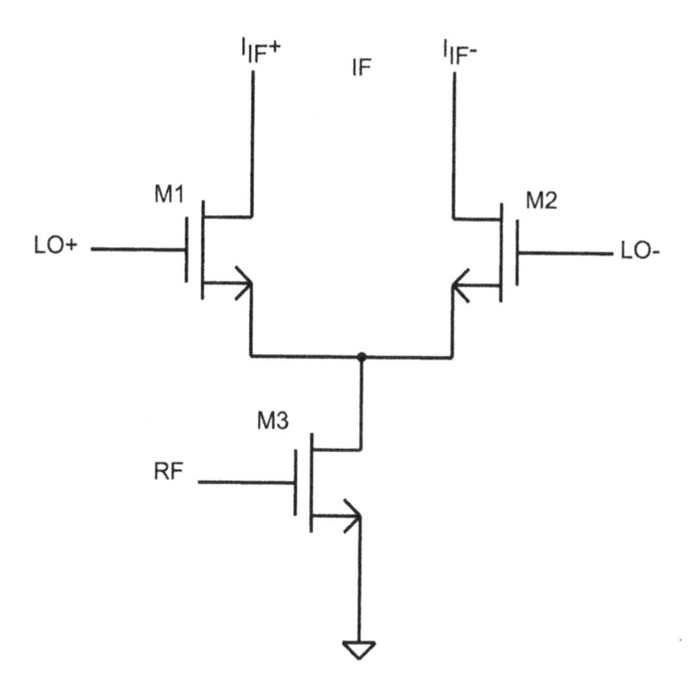

Figure 6.3. Gilbert mixer

terms such as $v_{RF}^3(t)$ and $v_{RF}^5(t)$. Those nonlinear terms distort the signal, i.e., compress the fundamental signals and generate intermodulation signals.

2. Gilbert mixer

Another popular type of mixer is the Gilbert cell mixer. Instead of commutating the RF signals in voltage, a Gilbert mixer commutates the RF signals in current [9].

Figure 6.3 shows an example of a Gilbert mixer. A transistor converts the RF input voltage into a current, and then a differential pair of transistors commutate the current to the complementary IF outputs each LO period. Because it doesn't need a large swing between the gates of the differential pair to commutate the current, the requirement of the LO drive is greatly reduced.

A Gilbert mixer provides better isolation between LO and RF than a passive mixer because there is no direct signal path from LO to RF. However, there is still LO leakage into the IF port through the parasitic capacitors between the gate and the drain of the switches in a simple Gilbert mixer as shown in Figure 6.3. A double-balanced Gilbert mixer solves this problem by coupling differential LO signals into the same IF output. As shown in Figure 6.4, each side of IF output is connected with two switches with 180 degree phased LO signals so

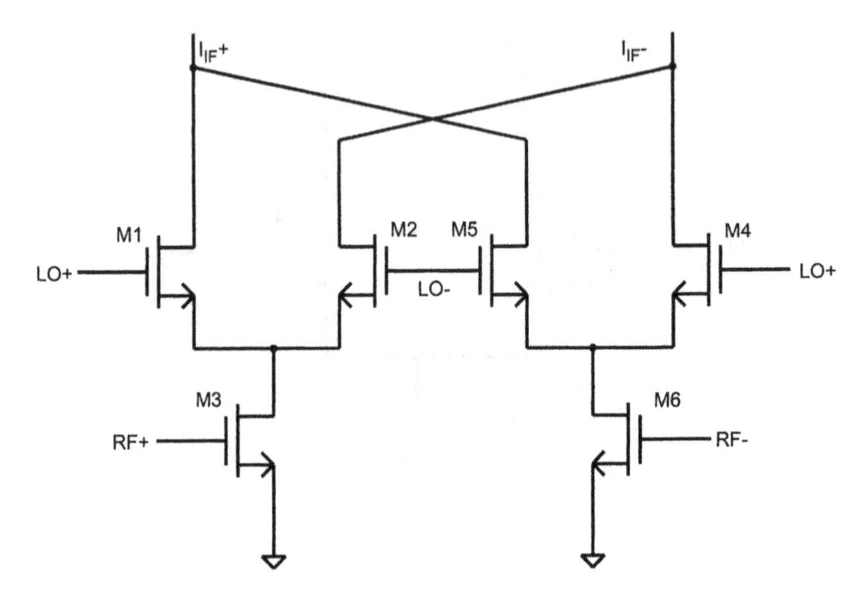

Figure 6.4. Double-balanced Gilbert mixer

that the LO leakage from the two switches cancels each other. Therefore, only the mixed products of RF and LO appear at the IF outputs.

In the following subsections conversion gain, noise and linearity of a double-balanced Gilbert mixer are discussed.

2.1 Conversion gain

As opposed to passive mixers, a Gilbert mixer can have a positive conversion gain. The conversion gain consists of three parts, the transconductance of the RF input MOS ($g_{m,rf}$), the switching gain/loss of the Gilbert cell (A_{sw}) and the output impedance (R_o). The voltage conversion gain is given by Equation 6.6.

$$A_v = g_{m,rf} R_o \cdot A_{sw} \tag{6.6}$$

Here, A_{sw} is a function of the shape and the amplitude of the LO drive and the overdrive voltage of the switching pair ($V_{od,sw}$). If the LO signal is a square wave with sharp rising and falling edges and its amplitude is larger than $V_{od,sw}$, i.e., the interval time of both the transistors of a switching pair being on is negligible, the switching gain is $2/\pi$, i.e., -3.9 dB. If the LO signal is a sinusoid with a amplitude V_{LO} much larger than $V_{od,sw}$, the switching gain is close to that using a square wave; therefore, the switching gain is close to

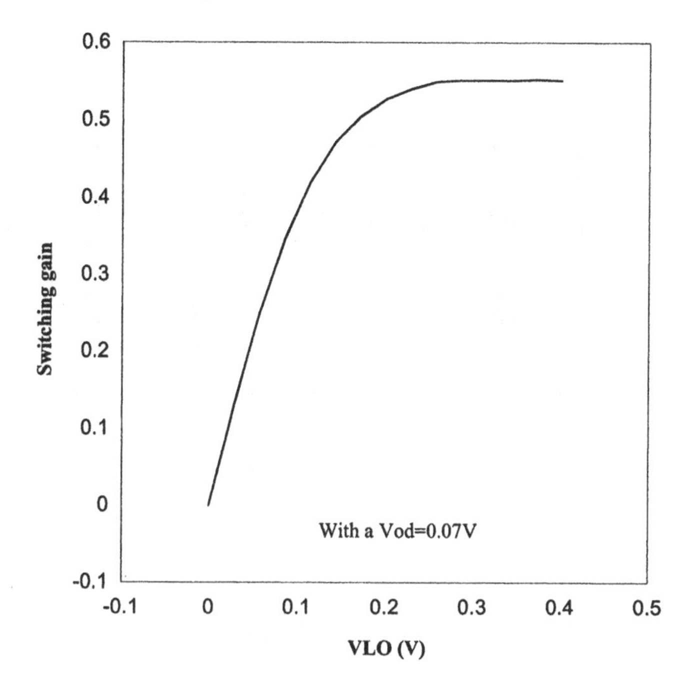

Figure 6.5. Switching gain curve of a double-balanced Gilbert mixer

$2/\pi$. Figure 6.5 shows the switching gain of a typical double-balanced Gilbert mixer. The switching gain is proportional to the LO amplitude when the LO amplitude is smaller than the over-drive voltage; and it is clamped to the point which is slightly less than $2/\pi$ due to the parasitic loss when the LO amplitude is sufficiently large.

The overdrive voltage of the switching transistor depends on the drain current of the RF input transistor and the size of the switching transistor. $V_{od,sw}$ can be estimated by a long channel device I-V equation as is shown in Equation 6.7.

$$V_{od,sw0} = \sqrt{\frac{I_{d,rf}}{\mu_n C_{ox} \frac{W}{L}}} \tag{6.7}$$

When the channel of the switching transistor become short enough so that the short channel equation has to be applied, $V_{od,sw}$ is then given by Equation 6.8, where ρ_0 gives the measure of velocity saturation effect as in Equation 6.9.

$$V_{od,sw} = V_{od,sw0} \left[\frac{\rho_0}{2} + \sqrt{1 + \left(\frac{\rho_0}{2}\right)^2} \right] \tag{6.8}$$

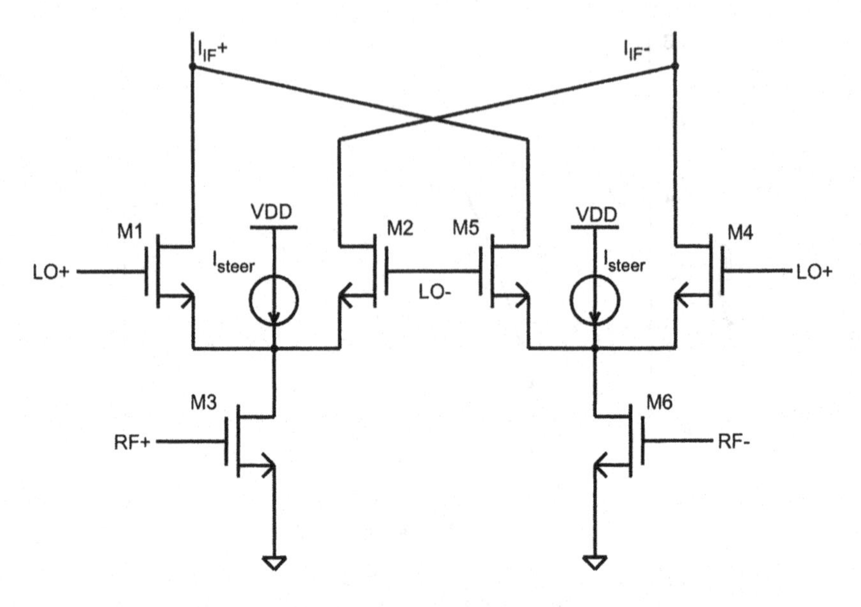

Figure 6.6. Double-balanced Gilbert mixer with DC current stealing

$$\rho_0 = \frac{V_{od,sw0}}{LE_{sat}} \tag{6.9}$$

Although a larger LO drive can provide a higher switching gain, too large a LO drive also degrades the conversion gain because even order harmonics are coupled into the common source of the differential pair, which is also connected to the drain of the RF input transistors. A large LO harmonic can reduce the drain voltage of the input transistor and finally push the transistor into the triode region.

Instead of increasing the LO drive, decreasing the over-drive voltage of the differential pair can increase the conversion gain. DC current stealing is often used to achieve a small over-drive voltage, as is shown in Figure 6.6. A current source is connected to the common source of the differential pair to steal part of DC current from the drain of the input transistor so that there is less DC current flowing through the differential pair and the over-drive voltage becomes smaller. Only the DC current at the output is reduced while the AC current which contains all the signals remains unchanged.

2.2 Noise

There are three major noise sources in a down-conversion mixer: the noise generated in the RF input transistors, the switching noise and the noise from the output loads.

The noise generated in the input transistors is primarily from two parts. First is the drain thermal noise, as is given by Equation 6.10. The input referred drain noise voltage of this source can also be given by Equation 6.11.

$$i_{d,noise}^2 = \frac{8}{3} kT g_{m,rf} \quad (6.10)$$

$$V_{in,thermalnoise}^2 = \frac{8}{3} \frac{kT}{g_{m,rf}} \quad (6.11)$$

The second noise source in the input transistor is the induced gate noise, whose input referred value is given by Equation 6.12, where ω is the operating angular frequency, C_{gs} is the gate-to-source capacitance and g_{d0} is the drain conductance when V_{ds} equals zero.

$$i_{g,noise}^2 = 4kT\delta g_g \quad (6.12)$$

where

$$g_g = \frac{\omega^2 C_{gs}^2}{5 g_{d0}} \quad (6.13)$$

The induced gate noise is partially correlated to the drain thermal noise. The noise figure of the mixer can be optimized by use of a proper input impedance. The detailed analysis is similar to LNA noise optimization which was discussed in Chapter 5.

The differential pair switches the RF current between the two transistors at LO frequency. It also contributes noise into the signal path. One noise contribution is from the switching loss and the other is from the noise on the LO signals. The signal-to-noise ratio decreases by the same amount as the switching loss, which in turn increases the noise figure of the mixer. The noise at the gate of the differential pair consists of the phase noise and the thermal noise on the LO signals and the induced gate noise. When the LO amplitude is much higher than the differential pair's over-drive voltage, i.e., the interval of both the transistors of the differential pair on is much smaller than the LO period, both the LO thermal noise and the induced gate noise have much less impact than the LO phase noise.

In conclusion, the Gilbert mixer can provide a smaller noise figure than a passive mixer. It can also have gain instead of loss, which together with LNA reduces the noise contribution from the IF stages, such as IF filters.

2.3 Linearity

The linearity of the Gilbert mixer is often limited by the transconductance of the RF input MOSFETs.

The linearity of the RF input transconductance can be expressed in term of IIV3 or IIP3. The IIV3 of a common source MOSFET is given by Equation 6.14 [48], where α is a measure of the channel velocity saturation and the mobility degradation.

$$IIV_3 = \sqrt{\frac{8}{3}\frac{1}{\alpha}V_{od}\left(1+\frac{1}{2}\alpha V_{od}\right)(1+\alpha V_{od})^2} \tag{6.14}$$

Note that the IIV3 increases with the over-drive voltage V_{od}. One method to increase the linearity of a Gilbert mixer without reducing its conversion gain is to increase the drain current of the RF input transistors, and then steal the unnecessary DC current away from the signal path, as is shown in Figure 6.6.

The switching devices don't contribute much distortion into the outputs. A Gilbert cell commutates current instead of voltage. When the LO drive is much larger than the over-drive voltage, the differential pair switches the current almost perfectly so that the conversion gain is constant over the input current. However, with such an abrupt current commutating the mixing products of RF signals and higher order LO harmonics are created at the mixer output. The frequencies of the output signals can be given generally by Equation 6.15, where m and n are both odd numbers, as $m = 1, 3, 5, ...$ and $n = 1, 3, 5, ...$ Usually there is a low-pass IF filter following the mixer, thereby eliminating all higher frequency products other than $|f_{RF} - f_{LO}|$. In a double-balanced Gilbert mixer all the even order harmonics of both the RF and LO are cancelled.

$$f_{IF} = |nf_{RF} \pm mf_{LO}| \tag{6.15}$$

3. Gilbert mixer with harmonic cancellation

The harmonic cancellation technique can be applied to a Gilbert mixer to improve its linearity. Figure 6.7 shows a Gilbert mixer design with harmonic cancellation. Since the input transistors are the major distortion contributors, the cancellation is focused on the input stage. The harmonic cancallation greatly improves the IIV3 but has little impact on the conversion gain and noise figure.

3.1 Conversion gain

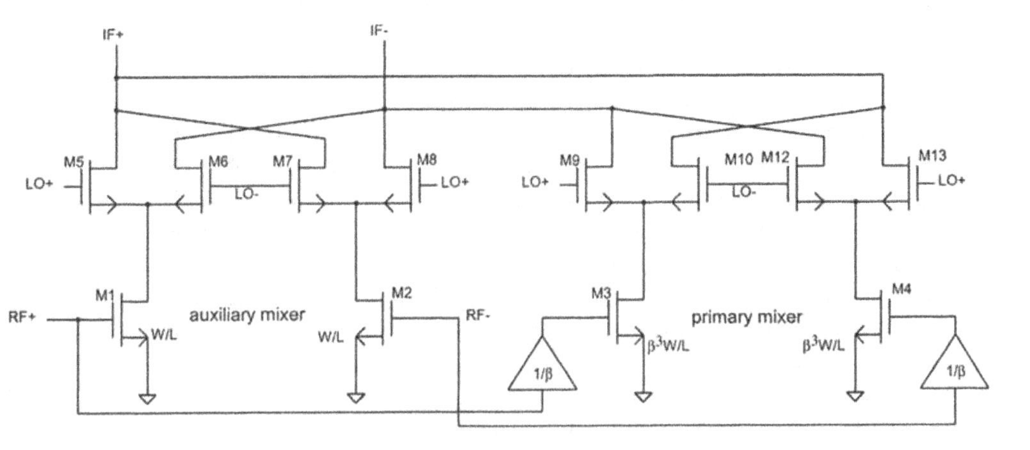

Figure 6.7. Double-balanced Gilbert mixer with harmonic cancellation

Harmonic cancellation improves the linearity of the down-conversion mixer, but also reduces the conversion gain. Compared to the traditional Gilbert mixer in Figure 6.4, the voltage gain reduction is given by Equation 6.16, where $A_{v,hcmxr}$ is the voltage conversion gain of the Gilert mixer with harmonic cancellation, and $A_{v,mxr}$ is the voltage conversion gain of the traditional Gilbert mixer. When β equals 2, the gain is reduced by 8.5 dB, which is a large reduction for a down-conversion mixer.

$$\frac{A_{v,hcmxr}}{A_{v,mxr}} = \frac{1}{\beta}\left(1 - \frac{1}{\beta^2}\right) \qquad (6.16)$$

However, in practice the reduction of the total gain of the LNA and the mixer is much less than 8.5 dB. Figure 6.8 shows a design of an LNA and a typical Gilbert mixer and Figure 6.9 shows a design of an LNA and a Gilbert mixer with harmonic cancellation. The linear loss stage is realized by series capacitors before the input transistors, i.e., the series capacitance satisfies Equation 6.17 to provide a voltage division of β.

$$C_{ac} = \frac{1}{\beta - 1} C_{gs} \qquad (6.17)$$

For a RF CMOS design the on-chip spiral inductors have a fairly constant quality factor. The voltage gain of the simple mixer with LNA in Figure 6.8 is given by Equation 6.18, where $g_{m,LNA}$ is the transconductance of LNA, $R_{o,LNA}$ is the output impedance of LNA, $g_{m,mixer}$ is the transconductance of the RF input stage of the mixer, A_{sw} is the switching gain of the mixer and the $R_{o,IF}$ is the output impedance of the mixer. In most cases $R_{o,LNA}$ is determined

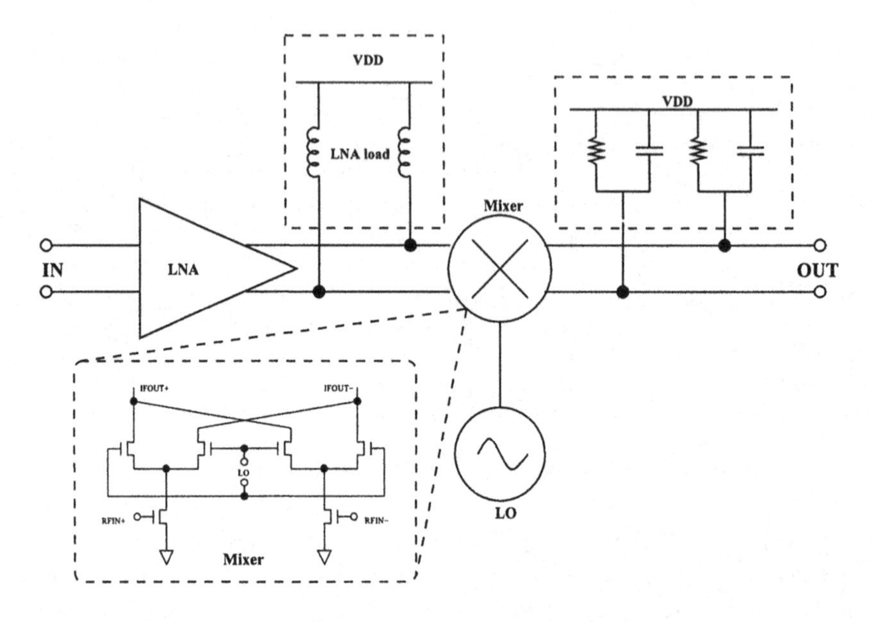

Figure 6.8. LNA and Gilbert mixer

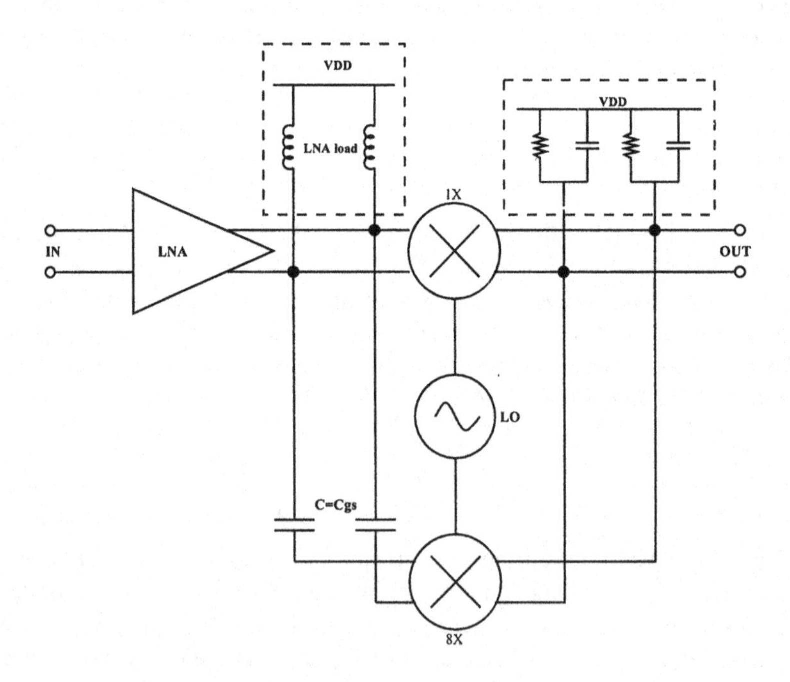

Figure 6.9. LNA and Gilbert mixer with harmonic cancellation

by the inductor value ($L_{o,LNA}$), the quality factor of the inductor (Q_L) and the resonant frequency (f_o).

$$\begin{aligned} Av_{LNA,mixer} &= g_{m,LNA} \cdot R_{o,LNA} \cdot g_{m,mixer} A_{sw} R_{o,IF} \\ &= g_{m,LNA} \cdot Q_L \omega_o L_{o,LNA} \cdot g_{m,mixer} A_{sw} R_{o,IF} \end{aligned} \quad (6.18)$$

When the circuit is running at the resonant frequency, i.e., $\omega_o^2 L_{o,LNA} C_{o,LNA} = 1$, the voltage gain of LNA and mixer is given by Equation 6.19. $C_{o,LNA}$ is the total capacitance at the outputs of LNA, which includes the C_{gs} of mixer, the parasitic diffusion capacitance, the parasitic capacitance of the inductor and the parasitic capacitance of the interconnects. In most cases the C_{gs} of the mixer transistor dominates the parasitic capacitance; therefore $C_{o,LNA} \simeq C_{gs}$.

$$\begin{aligned} Av_{LNA,mixer} &= g_{m,LNA} Q_L \frac{1}{\omega_o C_{o,LNA}} g_{m,mixer} A_{sw} R_{o,IF} \\ &\simeq g_{m,LNA} Q_L \frac{1}{\omega_o C_{gs}} g_{m,mixer} A_{sw} R_{o,IF} \end{aligned} \quad (6.19)$$

When harmonic cancellation is applied to the mixer, series capacitors are used before the primary input transistors so that the effective capacitor at the LNA output becomes smaller, as is given by Equation 6.20.

$$C_{o,LNA,hc} = \frac{1}{\beta}\left(1 + \frac{1}{\beta^2}\right) C_{gs} \quad (6.20)$$

The total voltage gain of LNA and the Gilbert mixer with harmonic cancellation is given by Equation 6.21.

$$Av_{LNA,mixer,hc} = \left(1 - \frac{1}{\beta^2}\right) g_{m,LNA} Q_L \frac{1}{\omega_o C_{o,LNA,hc}} g_{m,mixer} A_{sw} R_{o,IF} \quad (6.21)$$

The total gain reduction including LNA and mixer with harmonic cancellation is then given by Equation 6.22.

$$\frac{Av_{LNA,mixer,hc}}{Av_{LNA,mixer}} = \beta\left(1 - \frac{1}{\beta^2}\right)\left(\frac{1}{1 + \frac{1}{\beta^2}}\right) \quad (6.22)$$

When β equals 2, the total gain is reduced by 4.4 dB.

3.2 Improvement of dynamic range

A circuit's dynamic range is determined by its noise figure and linearity. The noise figure specifies the lowest receivable signals, and the IIP3 gives the maximum input power with the tolerable distortion. In an RF front-end of a receiver, usually the LNA dominates the noise figure and the mixer dominates the linearity. Equation 6.23 and Equation 6.24 give the total IIP3 and NF of LNA and mixer together respectively.

$$\frac{1}{IIP_{3,total}} = \frac{1}{IIP_{3,LNA}} + \frac{A_{LNA}}{IIP_{3,mixer}} + ... \quad (6.23)$$

$$NF_{total} = 1 + (NF_{LNA} - 1) + \frac{NF_{mixer} - 1}{A_{LNA}} + ... \quad (6.24)$$

Compared to a traditional Gilbert mixer, the Gilbert mixer with harmonic cancellation has a much higher IIP3.

As is shown in Figure 6.9, the RF input voltage is divided by β time before it is amplified by the primary mixer. The direct inputs are applied to the inputs of an auxiliary mixer, which has an input transistor size of $\frac{1}{\beta^3}$ time of that of the primary mixer. Both mixers have the same bias voltage so that the DC current of the auxiliary circuit is $\frac{1}{\beta^3}$ times that of the primary circuit. The LO switches are also scaled by the same ratio of β^3, i.e., the switches of the primary circuit is β^3 times that of the auxiliary circuit, so that the over-drive voltage of all the switches are the same. Therefore, the primary mixer has an input $\frac{1}{\beta}$ times that of the auxiliary mixer, but has a conversion transconductance of β^3 times of the auxiliary mixer, as is shown in Equation 6.25. Here, $g_{conv,pri}$ and $g_{conv,aux}$ are the conversion transconductance of the primary mixer and the auxiliary mixer, respectively; $g_{m,prim}$ and $g_{m,aux}$ are the transconductance of the input transistors of the primary mixer and the auxiliary mixer, respectively; W and L are the width and the length of the input transistors of the auxiliary mixer; V_{od} is the over-drive voltage of the input transistors; and α is given by Equation 6.26.

$$\begin{aligned} g_{conv,aux} &= A_{sw} g_{m,aux} \\ &= A_{sw} \mu_n C_{ox} \frac{W}{L} V_{od} \frac{1 + \frac{\alpha}{2} V_{od}}{(1 + \alpha V_{od})^2} \left(1 + a v_{in}^2\right) \\ g_{conv,pri} &= A_{sw} g_{m,prim} \\ &= A_{sw} \beta^3 \mu_n C_{ox} \frac{W}{L} V_{od} \frac{1 + \frac{\alpha}{2} V_{od}}{(1 + \alpha V_{od})^2} \left(1 + a v_{in}^2\right) \end{aligned}$$

$$= \beta^3 g_{conv,aux} \tag{6.25}$$

$$\alpha = \theta + \frac{\mu_0}{2v_{sat}L} \tag{6.26}$$

Since V_{od} is the same for the primary mixer and the auxiliary mixer, and the width of the input transistors of the primary mixer is β^3 times larger than that of the auxiliary mixer, the transconductance of the input transistors of the primary mixer is β^3 times larger than that of the auxiliary mixer. A_{sw} is the same for both the primary mixer and the auxiliary mixer because the over drive voltage of the differential pairs is the same and the LO swings on both the differential pairs are the same. The short channel transconductance equation is used with a 3rd order distortion.

When the two sets of currents join together at the IF output with an opposite phase, the 3rd order harmonic from both mixers can be completely cancelled if there is no other higher order distortion contributing into the 3rd harmonic. The total differential drain current is given by Equation 6.27.

$$\begin{aligned} i_{d,tot} &= \frac{1}{\beta} v_{in} g_{conv,pri}\left(\frac{1}{\beta} v_{in}\right) - v_{in} g_{conv,aux}(v_{in}) \\ &= \left(\beta^2 - 1\right) A_{sw}\mu_n C_{ox}\frac{W}{L} V_{od}\frac{1+\frac{\alpha}{2}V_{od}}{(1+\alpha V_{od})^2} v_{in} \end{aligned} \tag{6.27}$$

However, in practice such a cancellation is limited by the input signal matching and the gain matching between two mixers. In most cases 30 dB to 40 dB 3rd harmonic compression can be achieved in CMOS processes with 1% matching. Further details of the limits of the harmonic cancellation scheme are given in Chapter 4.

Thirty dB 3rd harmonic compression and 4.4 dB gain reduction give an increase of IIP3 of almost 20 dB. Assuming that the LNA gain is sufficient to suppress the noise contribution from mixer with and without harmonic balance, the dynamic range of LNA and mixer together can be improved by 20 dB with the harmonic cancellation technique.

3.3 An example of a receiver RF front-end with harmonic cancellation

A receiver RF front-end with harmonic cancellation was designed in the UMC 0.18 μm CMOS process. This RF front-end consists of an inductive source degenerated LNA and a double-balanced Gilbert mixer with harmonic cancellation. The complete schematic is shown in Figure 6.10.

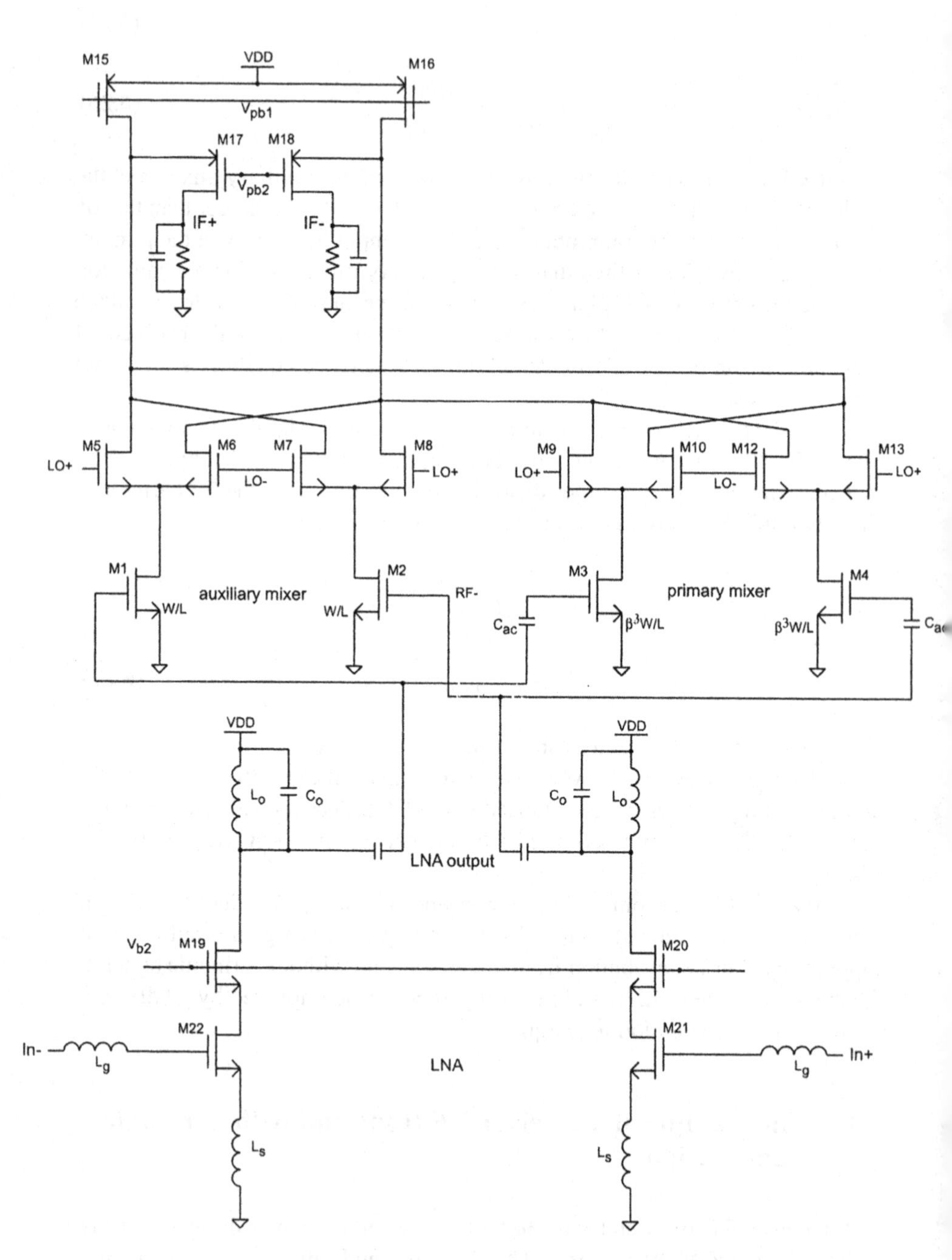

Figure 6.10. Complete schematic of LNA and Gilbert mixer with harmonic cancellation

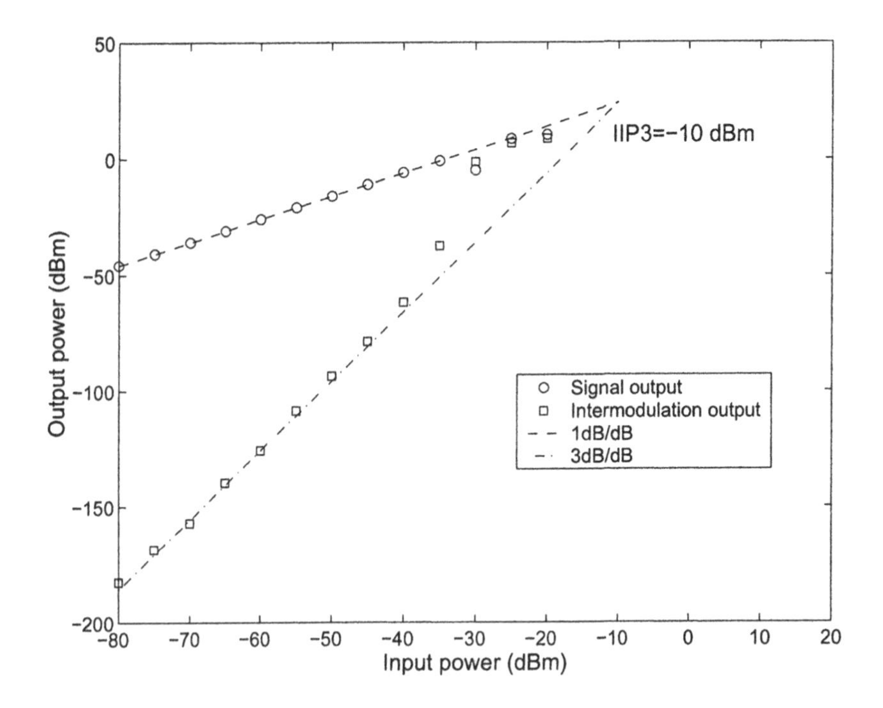

Figure 6.11. IIP3 curve of the traditional RF front-end

The RF front-end is operated at 5 GHz. The LNA is designed to provide 20 dB power gain with 2.5 dB NF and +2 dBm IIP3. The mixer with harmonic cancellation has 9 dB power gain, 14 dB NF and +32 dBm IIP3. Therefore, the RF front-end has a total power gain of 29 dB with 3.1 dB NF and +1.6 dBm IIP3. If the signal's band width is 20 MHz with requirement of 25 dB signal-to-noise ratio (SNR), the dynamic range of the RF front-end with harmonic cancellation is 50 dB, i.e., from -73 dBm to -23 dBm. Compared to a RF front-end with the same LNA but a traditional Gilber mixer, its dynamic range is increased by 12 dB. Figure 6.11 shows the IIP3 curve of the traditional RF front-end and Figure 6.12 shows the IIP3 curve of the RF front-end with harmonic cancellation. The simulation results are compared in Table 6.1.

4. Summary

Both the passive down-conversion mixer and the active Gilbert mixer are discussed in this chapter. Compared to the passive mixer the Gilbert mixer has a higher conversion gain and requires smaller LO drive amplitude, but consumes more DC power. The linearization technique of harmonic cancellation has been applied to a double-balanced Gilbert mixer. The linearity of the Gilbert mixer has been greatly improved. An example of a receiver RF front-end, consisting

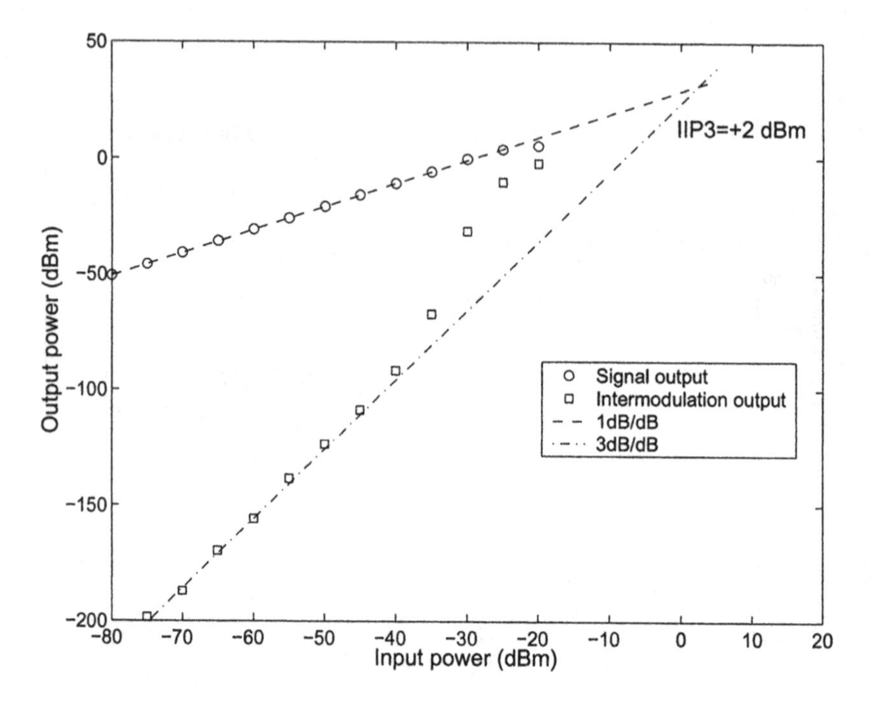

Figure 6.12. IIP3 curve of the RF front-end with harmonic cancellation

Performance Parameters	Traditional RF Front-end	RF Front-end with Harmonic Cancellation
Operating Frequency	5 GHz	5 GHz
Noise Figure (dB)	2.7	3.1
IIP3 (dBm)	-10.3	+1.6
Gain (dB)	34	29
Dynamic range	-73 dBm to -35 dBm	-73 dBm to -23 dBm
Supply Current (mA)	8.0	9.0

Table 6.1. Summary of simulation results

of a LNA and a mixer, was designed in the UMC 0.18 μm CMOS process. Simulations show a 12 dB improvement in the dynamic range of the entire RF front-end, which provides 29 dB power gain, 3.1 dB NF and +1.6 dBm IIP3.

Chapter 7

POWER AMPLIFIER DESIGN IN CMOS

The power amplifier (PA) is usually the last stage of a radio transmitter. Because of the propagation loss a radio has to transmit signals with high output power in order to cover a large area or a long distance. The power amplifier is used to amplify the signals with the radio frequency carriers and transmit them out into the air through an antenna.

There are two major categories of power amplifiers: linear amplifiers and constant envelope amplifiers. Linear amplifiers maintain all the phase and amplitude information of the signals during operation; but constant envelop amplifiers output signals with a constant amplitude. Constant envelope amplifiers are nonlinear amplifiers. They transmit signals with very coarse phase information and no amplitude information, i.e., binary phase information only. They are suitable for frequency modulation (FM) and binary phase modulation, such as BPSK. Because nonlinear power amplifiers don't need a large amount of DC power to maintain a constant gain over the entire operation range, they usually have a much higher power efficiency than linear power amplifiers.

However, for many wireless communication systems more complicated modulation schemes are used which requires a linear transfer function on both amplitude and phase, such as QPSK and QAM. A linear power amplifier has to be used for these applications. Power efficiency is a major concern in a linear power amplifier.

CMOS power amplifiers (PA) are used in modern radios primarily because they integrate well with digital circuits. Many linear power amplifiers have been designed for integrated radios [22, 20, 11, 27, 26, 41, 15, 29, 45, 43], and can be categorized as class A, class B, and class AB according to the duty cycle of the their drain currents [49].

This chapter discusses the traditional CMOS linear power amplifiers and also presents a new power amplifier with a parallel combination of a class A/AB

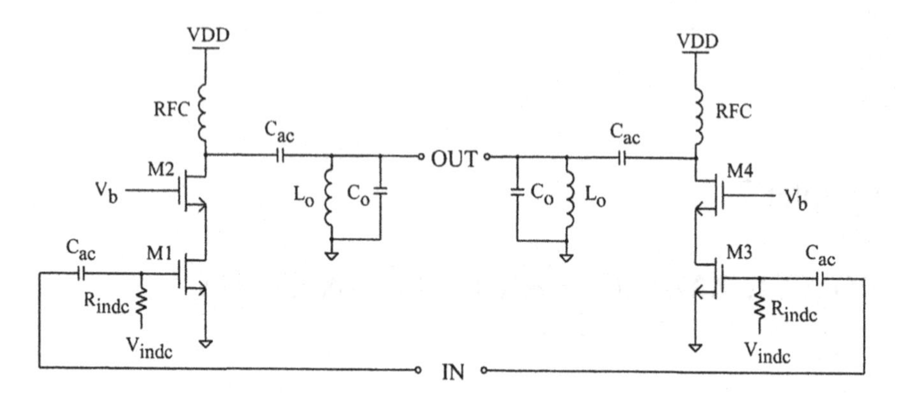

Figure 7.1. General CMOS power amplifier circuit

amplifier and a class B amplifier. This amplifier provides the high linearity of a class A amplifier while providing the PAE of a class B amplifier.

1. Traditional CMOS linear power amplifiers

Linear power amplifier can be categorized into different classes depending on the duty cycle of the their drain currents [49].

Figure 7.1 shows a simplified circuit diagram for a CMOS cascode power amplifier. Differential designs are often preferred for their improved power supply rejection ratio (PSRR) and suppression of even order harmonics. The input transistors M1 and M3 are biased at a fixed voltage, V_{indc}, through a pair of large valued resistors, R_{indc}. The cascode transistors M2 and M4 provide isolation between the input and output nodes. Capacitor C_{ac} is an AC coupling capacitor, RFC is the RF choke, and L_o and C_o are used for output matching. This basic topology can be made to operate in different modes, e.g., class A mode, class B mode, etc., by altering the input DC bias voltage V_{indc} to this amplifier.

1.1 Class A amplifier

Class A amplifiers are used where linear operation and high gain is critical. Figure 7.2 shows the typical drain current response of a class A amplifier when a sinusoid signal is applied. The drain current is always positive and sinusoid.

Gain

The transconductance for a class A amplifier, when the input is sufficiently small, is given by Equation 7.1, where, W and L are the input transistor device

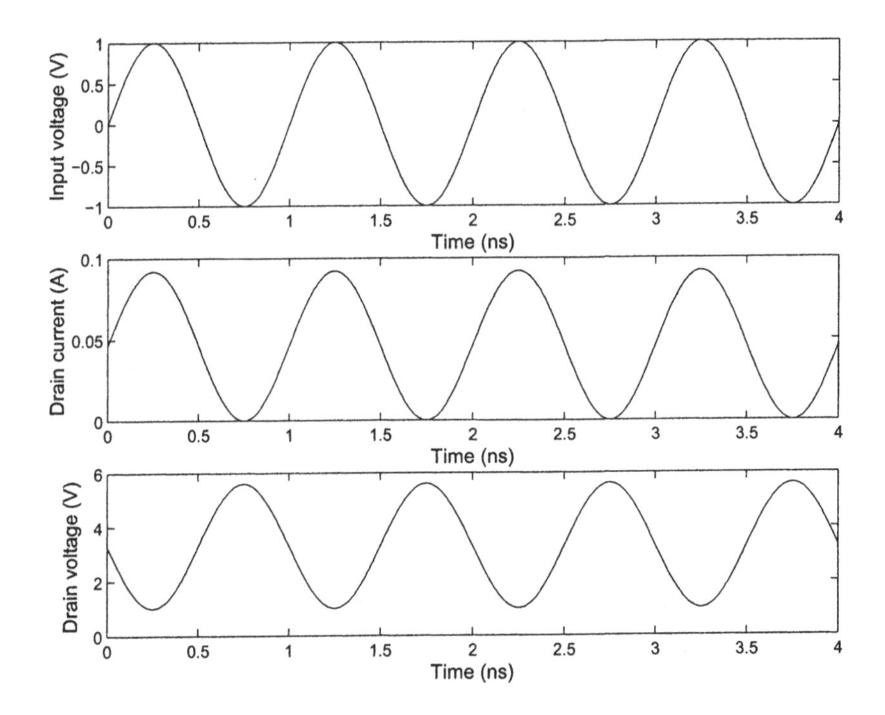

Figure 7.2. Drain current and voltage of class A amplifier

width and length, $V_{od} = V_{indc} - V_{th}$ is the over-drive voltage. The variable ρ_a, shown in Equation 7.2 is a measure of velocity saturation effect, and v_{sat} is the carrier saturation velocity.

$$g_{mA0} = \mu_n C_{ox} \frac{W}{L} V_{od} \frac{1 + \frac{1}{2}\rho_a}{(1 + \rho_a)^2} \tag{7.1}$$

$$\rho_a = \frac{\mu_n}{2\, v_{sat} \cdot L} V_{od} \tag{7.2}$$

Given the transconductance and both the input impedance (R_{in}) and the output impedance (R_L), the power gain of a class A amplifier can be calculated by Equation 7.3. A single stage class A power amplifier can easily provide more than 10 dB power gain.

$$A_p = g_{mA}^2 R_{in} R_L \tag{7.3}$$

Power efficiency

Drain efficiency (DE) is used to describe how much DC power the amplifier can translate into signal power, as is defined as Equation 7.4, where P_o is the output signal power and P_{DC} is the DC power consumption.

$$DE = \frac{P_o}{P_{DC}} \tag{7.4}$$

The DC current for a class A amplifier is given by Equation 7.5. To maintain class A operation throughout the input signal range such amplifiers need to be biased at a relatively high level. With the result that class A amplifiers consume significantly higher DC power than other amplifier topologies even for low level input signals.

$$I_{DCA} = \mu_n C_{ox} \frac{W}{L} V_{od}^2 \frac{1}{1+\rho_a} \tag{7.5}$$

With a known DC current and a fixed supply voltage V_{DD} the DC power consumption of a class A amplifier can be given as Equation 7.6

$$\begin{aligned} P_{DC} &= I_{DCA} \cdot V_{DD} \\ &= \mu_n C_{ox} \frac{W}{L} \frac{V_{od}^2 V_{DD}}{1+\rho_a} \end{aligned} \tag{7.6}$$

Power added efficiency (PAE) is also used to describe PA's performance, as is given by Equation 7.7, where P_{in} is the input signal power. A good PAE requires not only high drain efficiency but also sufficient power gain.

$$\begin{aligned} PAE &= \frac{P_o - P_{in}}{P_{DC}} \\ &= \left(1 - \frac{1}{A_p}\right) \cdot DE \end{aligned} \tag{7.7}$$

In theory the output voltage can swing as low as the ground voltage and up to twice the supply voltage ($2V_{DD}$) and the drain current can be as low as zero. In this scenario a class A amplifier has the highest efficiency. Its maximum drain efficiency is given by Equation 7.8 [49].

$$DE_{max} = \frac{\frac{1}{T}\int_0^T V_{d,max}(t) I_d(t) dt}{V_{DD} I_{DC}} = 50\% \tag{7.8}$$

In CMOS, the drain-to-source voltage (V_{ds}) has to be larger than the overdrive voltage (V_{od}) to keep the transistor in saturation. Therefore, the output signals cannot swing below V_{od} in order to keep its linear operation. The voltage level which sets the lowest output swing is referred as *knee voltage*. The knee voltage is one V_{od} for a signal FET class A PA, and two V_{od} for a cascode class A amplifier. With the result that the maximal drain efficiency is limited by the knee voltage, as is given by Equation 7.9.

$$DE_{max} = 50\% \cdot \left(1 - \frac{V_{knee}}{V_{DD}}\right) \tag{7.9}$$

Although the theoretical PAE of a class A amplifier can be as high as 50%, it is usually less than 30% in practice. Particularly class A amplifiers consume almost constant current over the entire operation range even at small inputs. Therefore, its PAE is very small at small inputs.

Linearity

The linearity of a class A amplifier in CMOS is primarily limited by its transconductor cell(s). Both the input signal level and the output signal level have an impact on the transconductance. When the input swings below the threshold voltage, the transconductance is compressed. Similarly, when the output swings below the knee voltage, the transconductance is also compressed.

A low order expansion of the nonlinear transconductance is given by Equation 7.10. The first order transconductance coefficient α_{A_0} is usually close to zero because it results in the second order harmonic at the output which is cancelled by the differential design. Further in Equation 7.10, g_{mA0} is the DC transconductance given by Equation 7.1, v_{in} is the normalized input signal amplitude as shown in Equation 7.11, and α_A, shown in Equation 7.12, is the 2nd order transconductance coefficient that results in the 3rd order harmonic of the input signal at the output. It can be seen from Equation 7.10 that the transconductance of the class A amplifier decreases with the input signal level.

$$g_{mA} = g_{mA0}\left(1 + \alpha_{A_0} v_{in} + \alpha_A v_{in}^2\right) \tag{7.10}$$

$$v_{in} = \frac{V_{in}}{V_{od}} \tag{7.11}$$

$$\alpha_A = -\frac{3\rho_a}{(1+\rho_a)^2(2+\rho_a)} \tag{7.12}$$

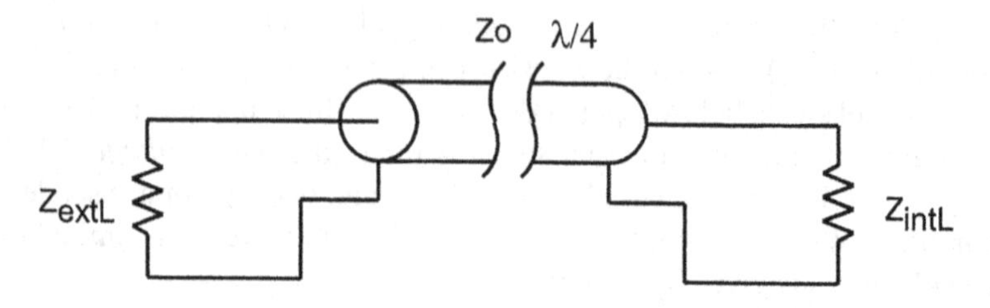

Figure 7.3. Impedance match by a quarter wave length transmission line

The linear operation range at the output of a power amplifier is limited by its knee voltage. A single FET amplifier is preferred for its lowest knee voltage. However, a single MOSFET amplifier has a significant miller capacitor between its gate and drain, which can provide a positive feedback and cause a serious stability problem. In most cases a cascode transistor is used to reduce the miller capacitance and provide a better isolation between the input gate and the output drain. As a result, the knee voltage of a cascode MOSFET amplifier is more than doubled compared to that of a single MOSFET amplifier.

One way to improve the output linear range is to use lower impedance loads. Without decreasing the knee voltage the output voltage swing is with the same constraint, but the output power increases with the lower load impedance. The relation of the output power and the load impedance is given by Equation 7.13.

$$P_o = \frac{V_o^2}{R_o} \tag{7.13}$$

In order to match to the standard external impedance, such as 50 Ω, a matching network is usually used to translate the external impedance to a lower impedance at the drain. A quarter wave length transmission line is one of the most popular impedance matching techniques, as is shown in Figure 7.3 [21]. Equation 7.14 gives the impedance translation of a quarter wave length transmission line, where Z_{extL} is the external load impedance, Z_{intL} is the internal load impedance and Z_o is the characteristic impedance of the transmission line.

$$Z_{intL} = \frac{Z_o^2}{Z_{extL}} \tag{7.14}$$

A quarter wave length transmission line is often used in PC board design and microwave design. However, because of the limitation on its physical size it is seldom used in RF IC designs. For example, a quarter wave length of a metal line in a typical CMOS process is about 7.5 mm at 5 GHz, which is longer than the size of most RF CMOS chips.

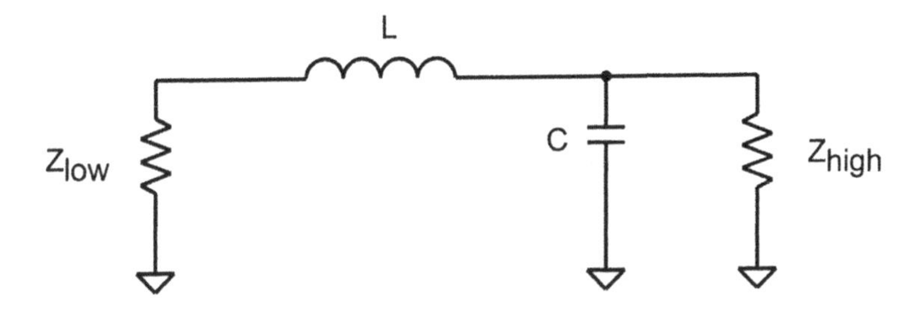

Figure 7.4. Impedance match by LC network

A LC network is often used in RF ICs to transform the impedances, as is shown in Figure 7.4 [49]. A high impedance can be translated into a low impedance, and vice versa. The translation is given by Equation 7.15, where Q_C is the quality factor of the capacitor C and the high impedance load Z_{high} at the resonant frequency of $\omega_o = \sqrt{\frac{1}{LC}}$, as is given by Equation 7.16.

$$Z_{low} = \frac{Z_{high}}{Q_C^2} \tag{7.15}$$

$$Q_C = \omega_o C Z_{high} \simeq \frac{\omega_o L}{Z_{low}} = Q_L \tag{7.16}$$

Single MOSFET amplifier with an inductive reverse isolation

With the help of an inductor to compensate the miller capacitor the single MOSFET amplifier can improve its stability and keep its small knee voltage. Figure 7.5 shows such an amplifier. An AC shunt inductor L_{gd} is used between drain and gate to resonate out the overlap capacitor so as to provide a better reverse isolation. A DC blocking capacitor is also used to separate the operating point of the drain from that of the gate. In order to prevent larger than unit gain feedback from the drain to the gate not only does the resonant frequency of the shunt LC tank of L_{gd} and C_{gd} need to be aligned with the output LC tank, but also its bandwidth needs to be larger than that of the output LC tank.

Figure 7.6 shows the comparison of measurement results of a cascode class A amplifier and a single MOSFET class A amplifier. Both amplifiers were fabricated in UMC 0.18 μm CMOS process. Although the single MOSFET amplifier has a lower gain than the cascode amplifier by almost 2 dB at the low input power, it has a larger linear operation range. For example, the output

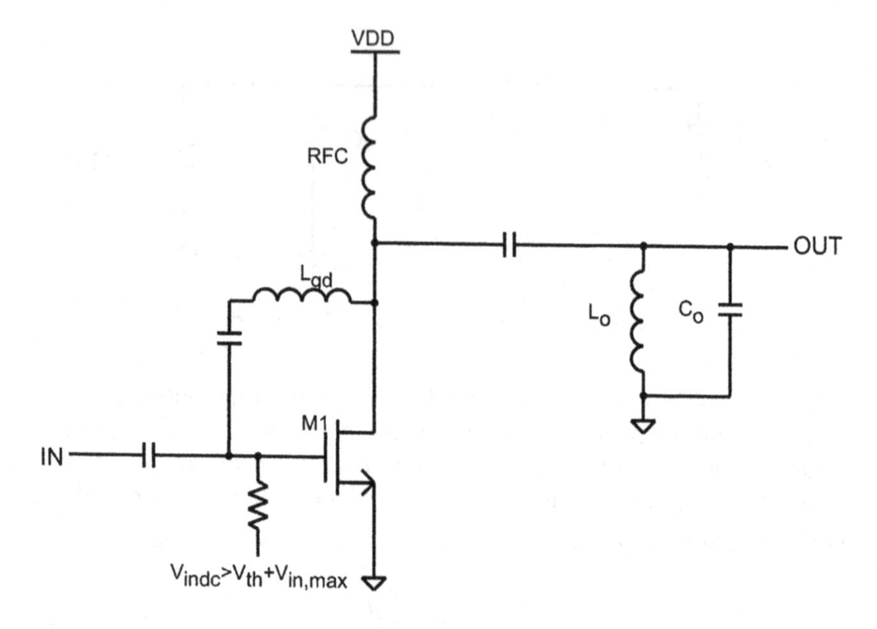

Figure 7.5. Single MOSFET amplifier with inductive reverse isolation

referred P1dB of the single MOSFET amplifier is +21 dBm, which is 3 dB higher than that of the cascode amplifier.

1.2 Class B amplifier

The drain current for a class B amplifier is on for only half the period of the sinusoidal input, as is shown in Figure 7.7. For a simple amplifier as shown in Figure 7.1 when its input transistors are biased at the threshold voltage, i.e., $V_{indc} = V_{th}$, it becomes a class B amplifier. A class B amplifier has the benefit of being able to self-adjust the DC power consumption depending on the input power level. Therefore, its power added efficiency (PAE) is much higher than that of a class A amplifier at small input levels. However, the gain of such amplifiers changes with the input signal level resulting in higher levels of distortion than the class A amplifiers.

Gain

Because the drain current of a class B amplifier is not constant, a large signal analysis has to be used to calculate the effective transconductance. The

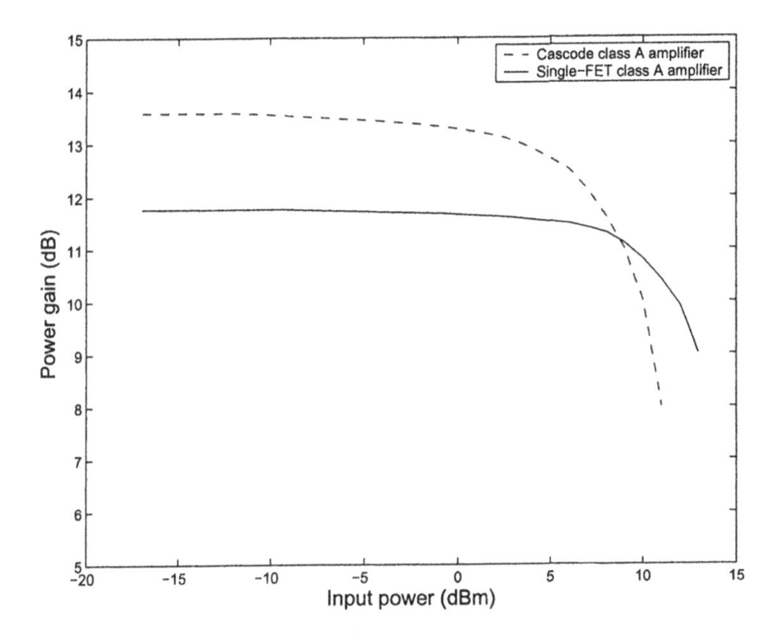

Figure 7.6. Measured gain curves of a single MOSFET class A amplifier and a cascode class A amplifier

transconductance of a CMOS class B amplifier can be calculated by using the expression in Equation 7.17, where, $I_{d,fund}$ is the drain current at the fundamental frequency and V_{in} is the amplitude of the input signal.

$$g_{mB} = \frac{I_{d,fund}}{V_{in}} \tag{7.17}$$

The drain current at the fundamental frequency, $I_{d,fund}$, of a CMOS class B amplifier with a sinusoid input of $V_{in} \cdot \sin(\omega t)$ is given by Equation 7.18,

$$\begin{aligned} I_{d,fund} &= \frac{2}{T}\int_0^{\frac{T}{2}} I_d(t)\sin(\omega t)dt \\ &= \frac{2}{T}\int_0^{\frac{T}{2}} \frac{1}{2}\mu_n C_{ox}\frac{W}{L}V_{in}^2\frac{\sin^2(\omega t)}{1+\rho_b\sin(\omega t)}\sin(\omega t)dt \\ &= \frac{1}{2}\mu_n C_{ox}\frac{W}{L}V_{in}^2[\frac{1}{2\rho_b} - \frac{2}{\pi}\frac{1}{\rho_b^2} + \frac{1}{\rho_b^3} \\ &\quad -\frac{4}{\pi}\frac{1}{\rho_b^3\sqrt{1-\rho_b^2}}\arctan\left(\sqrt{\frac{1-\rho_b}{1+\rho_b}}\right)] \end{aligned} \tag{7.18}$$

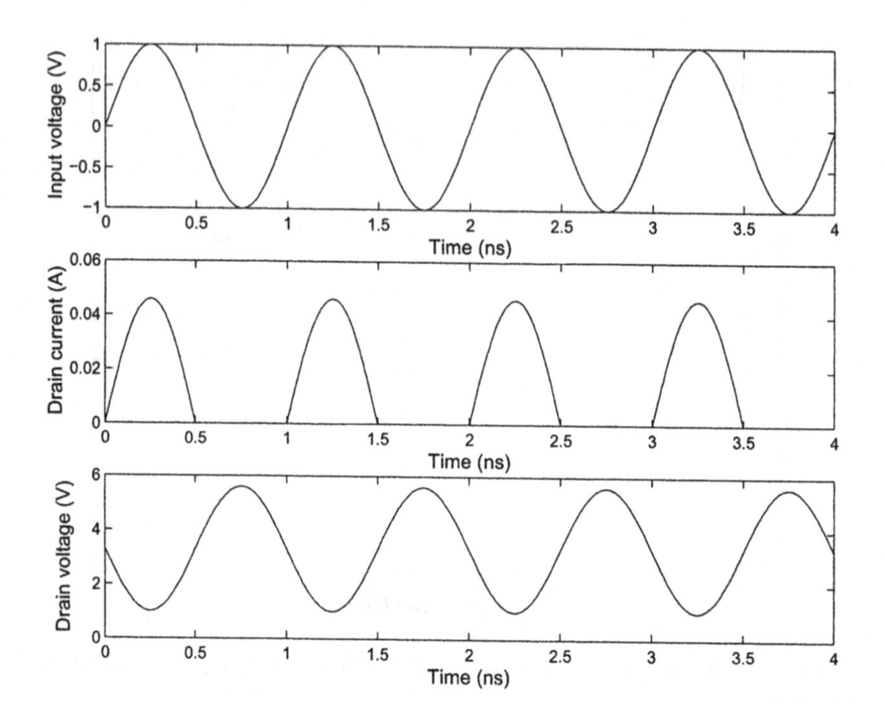

Figure 7.7. Drain current and voltage of class B amplifier

where ρ_b is given by

$$\rho_b = \frac{\mu_n}{2\,v_{sat} \cdot L} V_{in}$$

From Equations 7.17 and 7.18 we can derive an expression for the effective transconductance of a class B amplifier as shown in Equation 7.19.

$$g_{mB} = \frac{1}{2}\mu_n C_{ox}\frac{W}{L}V_{in}\left[\frac{1}{2\rho_b} - \frac{2}{\pi}\frac{1}{\rho_b^2} + \frac{1}{\rho_b^3} - \frac{4}{\pi}\frac{1}{\rho_b^3\sqrt{1-\rho_b^2}}\arctan\left(\sqrt{\frac{1-\rho_b}{1+\rho_b}}\right)\right. \quad (7.1$$

For very small inputs, the transconductance of the class B amplifier can be approximated by Equation 7.20. As can be seen in Equation 7.20 the transconductance increases with the input signal level until it reaches the compression point.

$$g_{mB}\mid_{V_{in}\to 0} = \frac{2}{3\pi}\mu_n C_{ox}\frac{W}{L}V_{in} \quad (7.20)$$

Same as other amplifiers, the power gain of a class B amplifier is given by Equation 7.21.

$$A_p = g_{mB}^2 R_{in} R_L \tag{7.21}$$

Power efficiency

With a sinusoidal input the drain efficiency of a class B amplifier is given by Equation 7.22, where I_{DCB} is the average drain current, which is given by Equation 7.23 and can be approximated by Equation 7.24 when the input is very small.

$$DE = \frac{\frac{1}{2} I_{d,fund} \cdot V_{d,fund}}{I_{DCB} \cdot V_{DD}} \tag{7.22}$$

The DC current increases with the input signal level. With a result that, a class B amplifier consumes much less DC current for small input signals than a class A amplifier.

$$\begin{aligned} I_{DCB} &= \frac{1}{T}\int_0^{\frac{T}{2}} I_d(t)dt \\ &= \frac{1}{2}\mu_n C_{ox}\frac{W}{L}V_{in}^2 \left[\frac{2}{\pi\rho_b} - \frac{1}{\rho_b^2} + \frac{4}{\pi\rho_b^3}\frac{1}{\sqrt{1-\rho_b^2}} \arctan\left(\sqrt{\frac{1-\rho_b}{1+\rho_b}}\right)\right] \end{aligned} \tag{7.23}$$

$$I_{DCB}\,|_{V_{in}\to 0} = \frac{1}{8}\mu_n C_{ox}\frac{W}{L}V_{in}^2 \tag{7.24}$$

The maximum drain efficiency is achieved when the output voltage swing reaches the maximum level, which is V_{DD} if the knee voltage is ignored. From Equation 7.18 and Equation 7.23 the maximum drain efficiency of a long channel CMOS class B amplifier is approximated by Equation 7.25.

$$DE_{max} \simeq \frac{8}{3\pi} \simeq 85\% \tag{7.25}$$

With the knowledge of the drain efficiency and the power gain, the power added efficiency is given by Equation 7.26.

$$PAE = DE \cdot \left(1 - \frac{1}{A_p}\right) \tag{7.26}$$

Although the drain efficiency of a CMOS class B amplifier can be as high as 85% in theory, the actual power efficiency is much lower for the limited output swing, the finite power gain and the extra power loss by the parasitics. In most cases PAE of a CMOS class B amplifier is below 50%, usually between 30% to 40%.

Linearity

A CMOS class B amplifier has a larger linear operation range than a CMOS class A amplifier. Unlike a CMOS class A amplifier, whose input linear range is limited by its DC over-drive voltage V_{od}, a CMOS class B amplifier can tolerate a much larger input swing as long as its output voltage is above the knee voltage. As a result its P1dB is higher than that of a CMOS class A amplifier.

However, as is given in Equation 7.20 and Equation 7.19, the gain of a CMOS class B amplifier is not completely linear. Instead of having a constant gain over the entire operation range as a CMOS class A amplifier does, the gain of a CMOS class B amplifier increases with input power until its output swing is close to the limit of the knee voltage. Furthermore, the close-to-threshold biasing forces the class B amplifier to work in weak inversion at low input power and results in more distortion in the gain curve.

1.3 Class AB amplifier

Class AB amplifiers have a drain current duty cycle that lies in between class A and class B amplifiers. For very small input signals the class AB amplifier essentially operates as a class A amplifier. But as the input signal level increases it naturally morphs into class AB mode. Most CMOS power amplifiers are operated in class AB mode for higher output power and better power added efficiency (PAE) [22, 11, 27, 26, 15, 45, 43, 29, 37]. Practical CMOS class AB amplifier can achieve up to 40% PAE at the maximum output power level. However, as the DC power consumption of such amplifiers is relatively constant over the entire operation range, its PAE becomes much smaller when the output power is lower than the maximum level. This becomes a serious problem for systems that have large dynamic-range (i.e., high peak-to-average ratio), such as multi-carrier OFDM (802.11a/802.11g) systems. For example, for the 802.11a standard 54 Mbps operation (64 QAM), the power amplifier needs to be backed

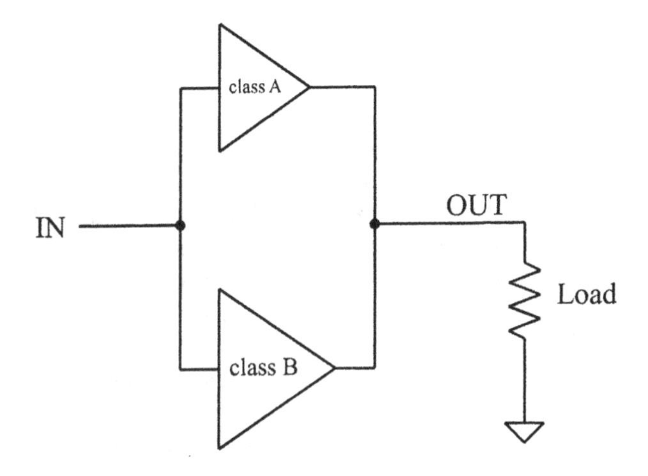

Figure 7.8. Conceptual block diagram for the parallel class A&B amplifier

off by 8 to 10 dB from its output P1dB point. With the result that during normal operation the PAE of the most class AB amplifiers effectively reduces to the 5% range.

2. CMOS parallel class A&B amplifier

A parallel combination of a class A and a class B amplifier can be used to improve both the linear operating range and the power efficiency simultaneously. A conceptual block diagram for such an amplifier is shown in Figure 7.8. As will be shown later the outputs from both the amplifiers are combined in the current domain with little overhead. As both amplifiers are operating in parallel this combined amplifier is called a *parallel class A&B amplifier.*

A parallel class A&B amplifier provides a better performance compromise for both large and small input levels in comparison to class A, class B and class AB amplifiers. With a proper ratio of sizes between the class A biased MOSFET and the class B biased MOSFET, a parallel class A&B amplifier has a larger linear range, a better power efficiency and a much lower DC power consumption.

2.1 Improvement of linear range

The transconductance for the parallel class A&B amplifier is a combination of the transconductance of the class A amplifier and the transconductance of the class B amplifier, as is given in Equation 7.27, where expressions for, g_{mA} and g_{mB} are given in Equation 7.1 and Equation 7.19, respectively.

$$g_{mA\&B} = g_{mA} + g_{mB} \tag{7.27}$$

From Equation 7.19 and Equation 7.20, it is clear that the effective transconductance and resulting power gain of a class B amplifier increases with increased input amplitude. However, the class A and/or class AB amplifiers starts to compress rapidly as the input signal level increases. In a parallel class A&B amplifier the class A amplifier is the primary transconductance contributor at low signal levels; however, the class B is the primary contributor at high signal levels. With the result that the class B amplifier can compensate for the compression of the class A amplifier when they are combined with the appropriate ratio. For example, Figure 7.9 shows the transconductances of a class A amplifier, a class B amplifier and parallel class A&B amplifiers with different combination ratios in the UMC 0.18μm CMOS process. Here, γ_{size} is the ratio of the class A transistor with respect to the class B transistor. In Figure 7.9 the class A transconductance decreases with the input signal level and the class B transconductance increases with the input signal level. The figure also includes three curves for different values of γ_{size} (1:1, 1:4 and 1:10). All values are plotted for an over-drive voltage of 0.5 V for the class A transistors. The variation in transconductance over the input signal range of 0 to 0.6 V is minimized when γ_{size} equals 1/4 (bold curve with diamonds (◇)). Increasing γ_{size} beyond this value as shown on the curve with the square (□) markers (γ_{size}=1/1) or decreasing it as shown on the curve with the triangle (△) markers (γ_{size}=1/10), results in a larger variation in the transconductance value.

Figure 7.10 shows the simulated transfer characteristics for the parallel class A&B amplifier. In this figure the solid line with the squares shows the input-output power transfer function for the parallel class A&B amplifier. The dashed line with the 'x's shows the contribution from the class A amplifier while the dashed line with the diamonds shows the contribution from the class B amplifier. As can be seen, at low input levels the class A amplifier contributes the majority of the gain as the gain of the class B is very low. As the input level increases the gain of the class B amplifier increases and its contribution to the overall gain increases proportionately. When the input level is sufficiently high, the class B amplifier provides the majority of the power gain and compensates for the gain compression of the class A amplifier.

In Figure 7.10 the class B amplifier compensates for the gain compression of the class A amplifier so that a higher P1dB for the parallel class A&B amplifier is expected in comparison to a class A amplifier. Additionally, since the class A amplifier does not have to provide significant power gain at larger inputs it can be designed to be smaller and biased at a lower power level, i.e., resulting in an improved PAE. To evaluate both the PAE and the linearity characteristics three amplifiers, class A, class B and parallel class A&B have been designed

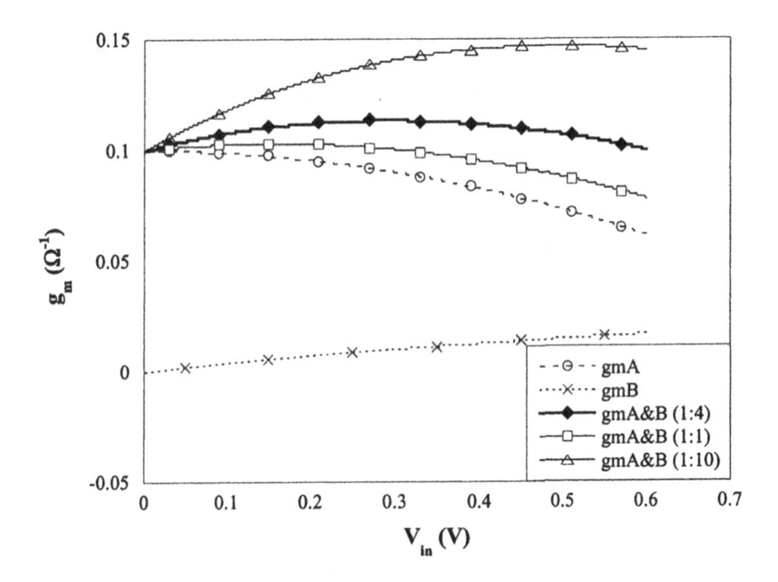

Figure 7.9. Transconductances for Class A, class B and parallel class A&B for different γ_{size}s

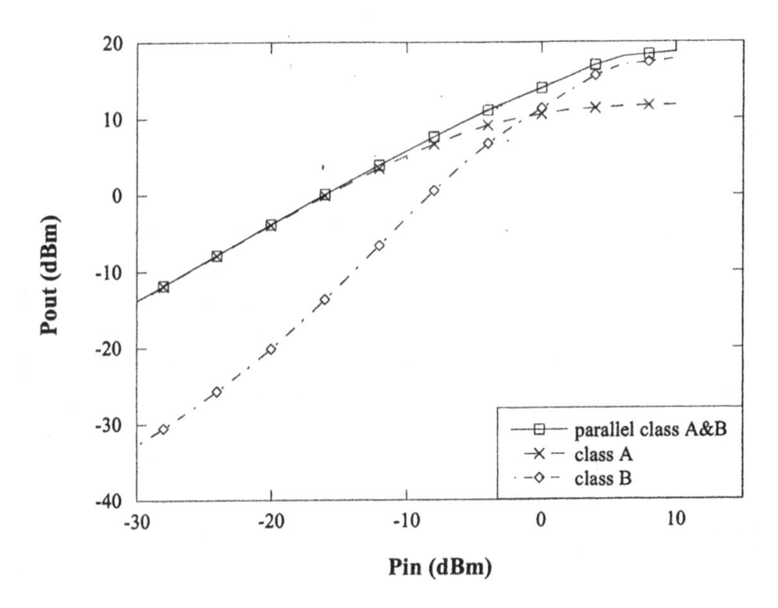

Figure 7.10. Parallel class A&B amplifier transfer characteristics

and simulated, all of which drive low impedance loads so that gain compression occurs at the inputs rather than at the outputs.

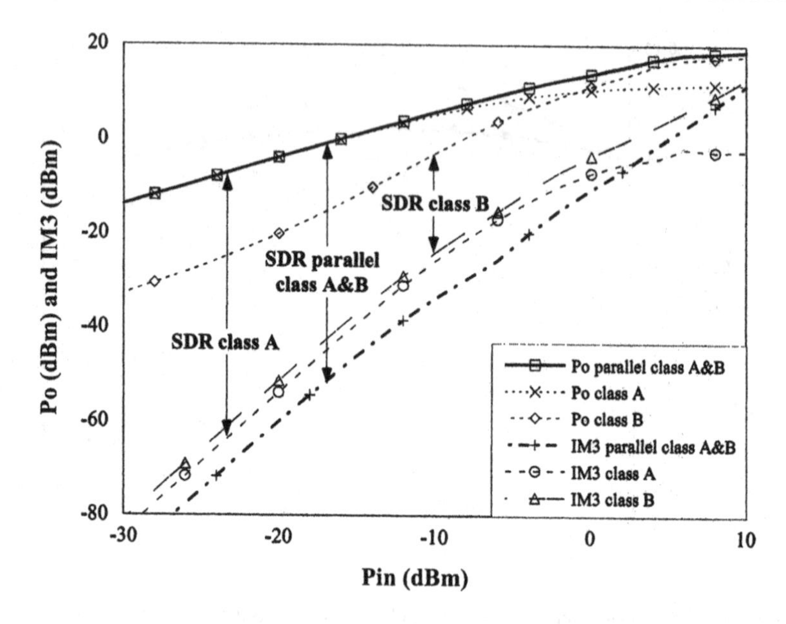

Figure 7.11. Comparison of intermodulation products

The linearity characteristics are shown in Figure 7.11 with both the fundamental and intermodulation signals (IM3) for the three amplifiers. Also marked on the figure is the signal-to-distortion ratio (SDR) for all three amplifiers. In the parallel class A&B amplifier the class A contributes to the majority of the gain at low signal levels so that the linearity is similar to a class A amplifier. Interestingly, the IM3 products are 5 to 7 dB lower than even the class A amplifier. This is because the class B amplifier partially compensates for the distortion caused by the class A amplifier. At high signal levels the class B amplifier contributes the majority of the output power, at which point the total distortion of the parallel class A&B amplifier is dominated by the class B part and is larger than that of the class A amplifier. Multicarrier systems such as 802.11a/802.11g are very sensitive to power amplifier distortion. In particular, 54 Mbps operation for either of these standards requires a minimum SDR of 25 dB. For example, the higher linearity of the parallel class A&B PA allows it to transmit 54 Mbps signals at 10 dBm output power while the class A amplifier with a SDR of 19 dB would only be able to support a 36 Mbps stream at the same output power.

Because the gate-to-source capacitance of the class B biased MOSFET varies over the gate-to-source voltage, the input impedance of the parallel class A&B amplifier depends on the input voltage. In order to reduce or eliminate the signal distortion caused by the input impedance change, a voltage source rather than a current source is preferred as the input signal source. For example, if a source

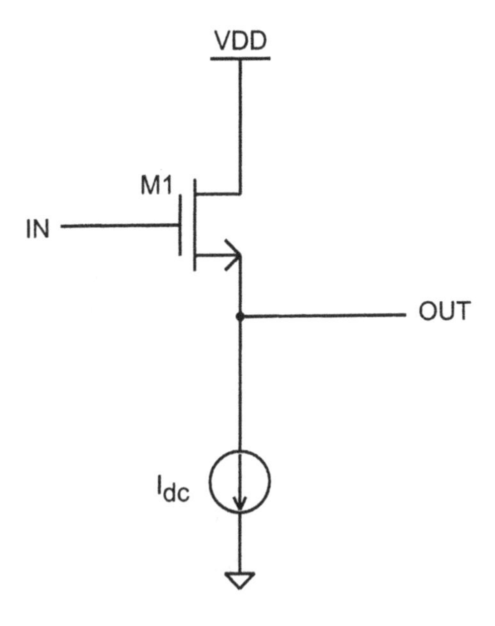

Figure 7.12. Source follower

follower as shown in Figure 7.12 is used to drive the power amplifier, the input voltage to the power amplifier is then independent of the gate capacitance.

2.2 Power efficiency improvement

The class A&B amplifier has a higher PAE than the class A amplifier. The DC current consumption of the parallel class A&B amplifier is given by Equation 7.28, where I_{DCA} and I_{DCB} are given by Equation 7.5 and Equation 7.23, respectively. The class A transconductor consumes the majority of the DC current for small inputs and the class B transconductors dominates the DC current consumption for large inputs.

$$I_{DCA\&B} = I_{DCA} + I_{DCB} \tag{7.28}$$

Figure 7.13 shows the power added efficiency (PAE) for the three power amplifiers vs. input signal power. The PAE of the class A amplifier is lower than both the class B amplifier and the parallel class A&B amplifier due to the higher DC bias. Because the class A part and the class B part of the parallel class A&B amplifier dominate the gain and the DC current at the different input power ranges, the PAE of the parallel class A&B is almost the same as that of

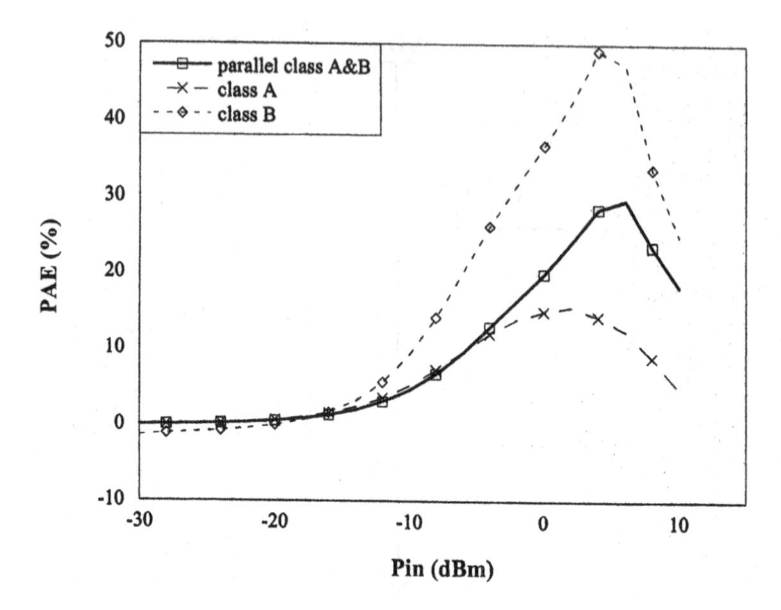

Figure 7.13. Comparison of PAE

the class A amplifier at the low inputs and close to that of the class B at the high inputs. The parallel class A&B amplifier gives the best compromise of the gain, the output power and the power added efficiency.

For these graphs a low output impedance is used to ensure that the power amplifier compresses at the input rather than at the output. However, in practice power amplifier are usually limited at both the input and output. If compression occurs at the output then the mode of operation of the transistor will have limited impact on the performance. Therefore, in practice the difference between the parallel class A&B amplifier and the class A/AB amplifiers maybe not be as large as shown in Figure 7.11 and Figure 7.13. But the improvements offered by the combined structure is still significant as will be seen via experimental results in the next section.

2.3 An example of parallel class A&B power amplifier

A prototype design of the parallel class A&B amplifier has been fabricated in the UMC 0.18μm RF CMOS process [53] and its performance is compared with other state-of-art CMOS power amplifiers in this section. A simplified circuit diagram of the differential parallel class A&B power amplifier design is shown in Figure 7.14.

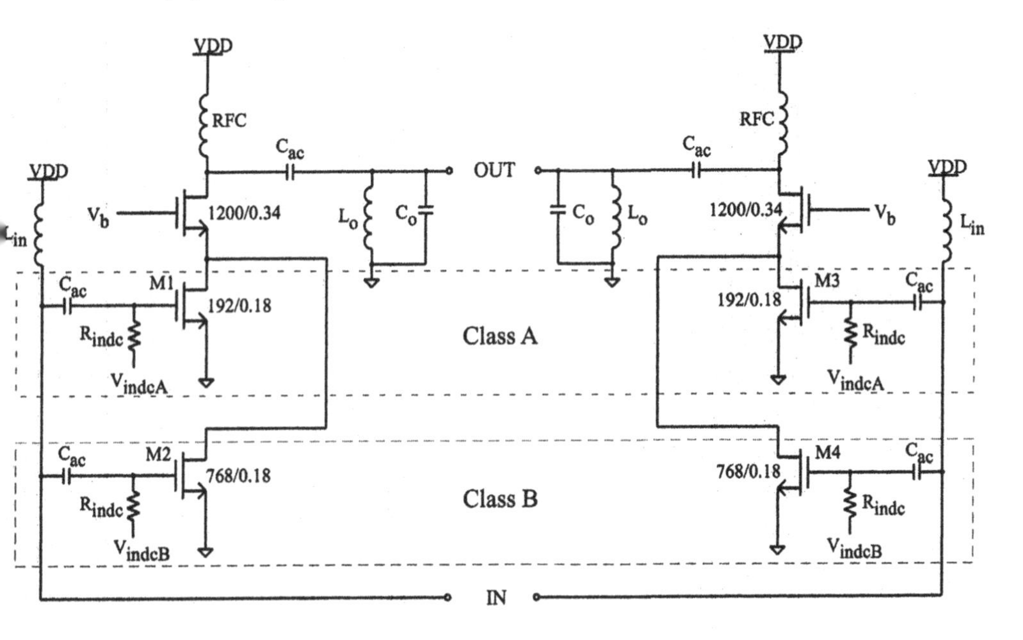

Figure 7.14. Circuit schematic for the CMOS parallel class A&B amplifier

Shunt inductors L_{in} are used to match the inputs, and the inductors RFC are used as RF chokes to prevent coupling of the RF signal to the power supplies. An additional pair of inductors L_o is used to match the output port. The input transistors M2 and M4 have an aspect ratio of (768/0.18) and are biased at the edge of the threshold voltage via the two large value resistors, i.e., $V_{indcB} \sim V_{th}$. These two transistors form the trasconductors for the class B amplifier. The other two input transistors M1 and M3 have an aspect ratio of (192/0.18) and are biased well above the threshold voltage so that they operate in saturation for the majority of the input range, i.e., $V_{indcA} \sim V_{th} + V_{od,max}$. These two transistors form the trasconductors for the class A amplifier. The size of the class A transistors M2 and M4 are only one quarter the size of the class B transistors M1 and M3. This is because the class B transistors do the "heavy lifting" at large input signals. The cascode transistors have an aspect ratio of (1200/0.34). These transistors have a thicker gate oxide with a higher breakdown voltage than the input transistors and are able to handle the large voltage swings at the outputs. A microphotograph of the fabricated power amplifier is shown in Figure 7.15.

The measured output power and PAE results for the parallel class A&B amplifier are shown in Figure 7.16. The power gain within the linear range is 12 dB, the maximum output power is over +22 dBm and the maximum PAE is

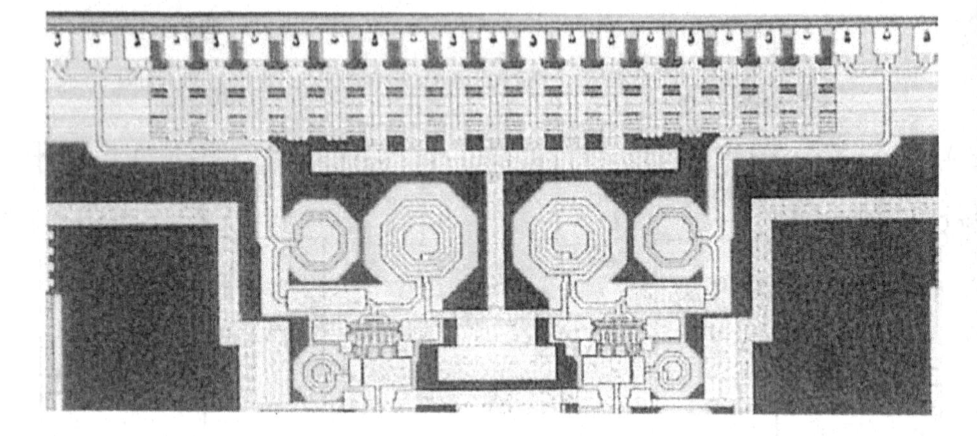

Figure 7.15. Die microphotograph of the CMOS parallel class A&B amplifier

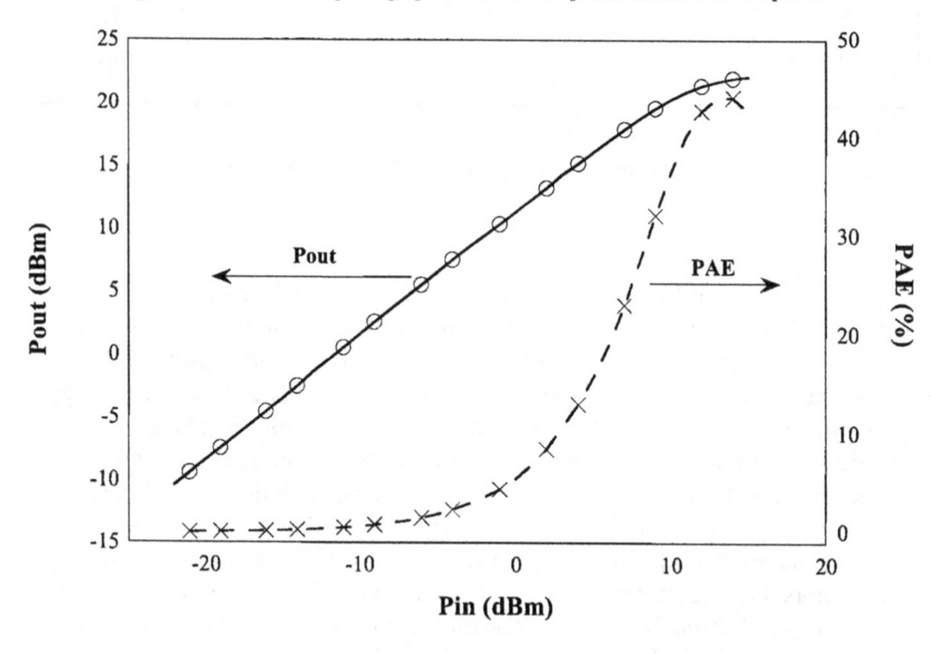

Figure 7.16. CMOS parallel class A&B amplifier measurement results

over 44%. The amplifier reaches its P1dB at an output level of +20.5 dBm and the PAE at P1dB is 36%.

Table 7.1 summarizes the measured results for the prototype amplifier and compares it with other CMOS class AB power amplifiers [11, 43, 27]. The new amplifier has a higher power added efficiency. This is particularly visible when the amplifiers are operated in their linear operating range, i.e., 4 dB to 8 dB

Design	A_p (dB)	$Pout_{max}$ (dBm)	PAE (%) @			
					Back-off	
			$Pout_{max}$	P1dB	4dB	8dB
This work 0.18 μm	12	22	44	36	18	8
Ballweber 0.6 μm [11]	5	19	30	26	14	6
Giry 0.35 μm [27]	24.6	23.5	35	24	13	6
Sowlati 0.18 μm [43]	36	23	42	18	8	4

Table 7.1. Comparison of measured results

back-off, that is usually mandated by other system requirements. For example at 8 dB back-off the PAE of the parallel class A&B amplifier is over 50% higher than the other designs. This can result in significant reduction in DC power consumption for similar RF output power levels.

3. Summary

Different classes of power amplifiers in CMOS have been discussed. Although the class A amplifier provides a greater linearity compared to other traditional power amplifiers, it consumes a large amount of DC power even at low input levels. The class B amplifier has a better power added efficiency than the class A amplifier, especially at low input power. However, the class B amplifier introduces significant nonlinearity.

A new power amplifier with a parallel class A&B structure has been presented. This new amplifier provides superior performance in terms of both linearity and power efficiency in comparison to previous designs. Measurement results show 12 dB power gain (single stage), 22 dBm output power and more than 44% PAE. More importantly this circuit uses significantly less DC power at 4 or 8 dB back-off in comparison to other class AB amplifiers.

Chapter 8

CONCLUSIONS

In this book we presented several linearization techniques for RF ICs, focusing the most critical RF circuits in both the receiver and the transmitter.

Some prior art transconductors have been discussed in Chapter 3. Although these transconductors improves the linear operation range at either the input or the output, none of these techniques can be easily used in RF ICs for their speed and stability concerns.

Chapter 4 presents a new linearization technique of harmonic cancellation. It uses an auxiliary circuit to cancel the unwanted harmonics from the primary circuit. Because it is a feed-forward technique and there is no need of any *a prior* knowledge of the signal harmonics of the primary circuit, it is a preferable linearization technique for RF ICs. With a CMOS process with 1% matching this technique can compress the signal distortion at the output by up to 40 dB. The auxiliary circuit doesn't impact the speed of the primary circuit and has a much smaller size than the primary circuit. Therefore, there is no significant trade-off of the power consumption, gain and noise.

To verify the linearization technique of harmonic cancellation, a low noise amplifier (LNA) has been designed and tested. Chapter 5 discusses the design details and gives the results. The LNA with harmonics cancellation has a 15 dB gain, 2.8 dB NF and +18 dBm IIP3. Compared to the same LNA but without harmonic cancellation, the IIP3 of the linearized LNA has been improved by 13 dB.

The harmonic cancellation linearization technique can also be applied to mixers. Chapter 6 discusses the design of a receiver RF front-end of a LNA and a down-conversion mixer with harmonic cancellation. The linearity of the mixer is improved by the same amount as the other direct amplification circuits. Because the LNA sets the noise figure of the entire front end and it is not impacted by the harmonic cancellation mixer, the overall dynamic range of the front end is greatly improved. Simulation results show more than a 12 dB improvement of the dynamic range with only 12% increase on power consumption.

Another linearization techniques have been used to improve the linear operation range of power amplifiers. As shown in Chapter 7 a single FET power amplifier with improved isolation between drain and gate provides a higher output power in linear range by 3 dB than a cascode CMOS class A amplifier. A new class of CMOS power amplifiers is also presented to improve the linear operation range and save DC power consumption. Because this class of amplifiers consists of a class A biased MOSFET and a class B biased MOSFET in parallel as its input transconductors, they are named as parallel class A&B amplifiers. Compared to the class A CMOS amplifier the parallel class A&B amplifier not only increased the linear operation range by 3 dB but also reduces the DC power consumption by more than 50%. The measurement results show an output power of +22 dBm with a PAE more than 44%.

References

[1] A. N. Karanicolas. "A 2.7-V 900-MHz CMOS LNA and Mixer". *IEEE Journal of Solid-State Circuits*, 31:1939–1944, December 1996.

[2] A. Nedungadi and T. R. Viswanathan. "Design of Linear CMOS Transconductance Elements". *IEEE Transaction on Circuits and System*, 31:891–894, October 1984.

[3] A. R. Shahani, D. K. Shaeffer and T. H. Lee. "A 12-mW Wide Dynamic Range CMOS Front-End for a Portable GPS Receiver". *IEEE Journal of Solid-State Circuits*, 32:2061–2070, December 1997.

[4] A.-S. Porret, T. Melly and C. C. Enz. "Design of High-Q Varactors for Low-Power Wireless Applications Using a Standard CMOS Process". In *IEEE Custom Integrated Circuits Conference*, pages 641–644, May 1999.

[5] Aldert Van Der Ziel. "Noise in Solid-State Devices and Lasers". In *Proceedings of IEEE*, volume 58(8), pages 1178–1206, August 1970.

[6] B. A. Floyd, J. Mehta, C. Gamero and K. O. Kenneth. "A 900-MHz, 0.8-μm CMOS Low Noise Amplifier with 1.2-dB Noise Figure". In *IEEE Custom Integrated Circuits Conference*, pages 661–664, 1999.

[7] B. Ray, T. Manku, R. D. Beards, J. J. Nisbet and W. Kung. "A Highly Linear Bipolar 1V Folded Cascode 1.9 GHz Low Noise Amplifier". In *Proceedings of Bipolar/BiCMOS Circuits and Technology Meeting*, pages 157–160, 1999.

[8] Ban-Leong Ooi, Dao-Xian Xu, Pang-Shyan Kooi, and Fu-Jiang Lin. "An Improved Prediction of Series Resistance in Spiral Inductor Modeling With Eddy-Current Effect". *IEEE Transaction on Microwave Theory and Techniques*, 50:2202–2206, September 2002.

[9] Barrie Gilbert. "A Precise Four-Quadrant Multiplier with Sub-nanosecond Response". *IEEE Journal of Solid-State Circuits*, SC-3(4):365–373, December 1968.

[10] Behzad Razavi. *RF Microelectronics*. Prentice Hall PTR, 1998.

[11] Brian Ballweber, Ravi Gupta and David J. Allstot. "Fully-Integrated CMOS RF Amplifier". In *IEEE International Solid-State Circuits Conference*, pages 72–448, 1999.

[12] C. Patrick Yue and S. Simon Wong. "On-Chip Spiral Inductors with Patterned Ground Shields for Si-Based RFIC's". In *VLSI Circuits Symposium Digest of Technical Papers*, pages 85–86, 1997.

[13] C. Patrick Yue, and S. Simon Wong. "Physical Modeling of Spiral Inductors on Silicon". *IEEE Journal of of Solid-State Circuits*, 47(3), 2000.

[14] C. S. Kim, M. Park, C. Kim, Y. C. Hyeon, H. K. Yu, K. Lee and K. S. Nam. "A Fully Integrated 1.9-GHz CMOS Low-Noise Amplifier". *IEEE Microwave and Guided Wave Letters*, 8:293–295, August 1998.

[15] D. Heo, A. Sutono, E. Chen, E.Gebara, S. Yoo, Y. Suh, J. Laskar, E. Dalton and E. M. Tentzeris. "A High Efficiency 0.25-μm CMOS PA with LTCC Multi-layer High-Q Integrated Passive for 2.4 GHz ISM Bnad". In *IEEE MTT-S International Microwave Symposium*, pages 915–918, 2001.

[16] D. K. Shaeffer, A. R. Shahani, S. S. Mohan, H. Samavati, H. R. Rategh, M. M. Hershenson, M. Xu, C. P. Yue, D. J. Eddleman and T. H. Lee. "A 115-mW, 0.5-um CMOS GPS Receiver with Wide Dynamic-Range Active Filters". *IEEE Journal of Solid-State Circuits*, 33:2219–2231, December 1998.

[17] D. K. Shaeffer and T. H. Lee. "A 1.5-V, 1.5-GHz CMOS Low Noise Amplifier". *IEEE Journal of Solid-State Circuits*, 32:745–759, May 1997.

[18] D. R. Welland, S. M. Phillip, KaY Leung, G. T. Tuttle, S. T. Dupuie, D. R. Holberg, R. V. Jack, N. S. Sooch, K. D. Anderson, A. J. Armstrong, R. T. Behrens, W. G. Bliss, T. O. Dudley, W. R. Forland, N. Glover and L. D. King. "A Digital Read/Write Channel with EEPR4 Detection". In *IEEE International Solid-State Circuits Conference*, pages 276–277, 1994.

[19] David A. Johns and Ken Martin. *Analog Integrated Circuit Design*. John Wiley & Sons, Inc., New York, 1996.

[20] David K. Su and William J. McFarland. "An IC for Linearizing RF Power Amplifiers Using Envelope Elimination and Restoration". *IEEE Journal of of Solid-State Circuits*, 33:2252–2258, December 1998.

[21] David M. Pozar. *Microwave Engineering*. John Wiley & Sons, Inc., 1998.

[22] David Su and William McFarland. "A 2.5-V, 1-W Monolithic CMOS RF Power Amplifier". In *IEEE Custom Integrated Circuits Conference*, pages 189–192, 1997.

[23] Donald D. Weiner and John F. Spina. *Sinusoidal analysis and modeling of weakly nonlinear circuits*. Van Nostrand Reinhold Company, New York, 1980.

[24] F. Krummenacher and N. Joehl. "A 4-MHz CMOS Continuous-Time Filter with On-Chip Automatic Tuning". *IEEE Journal of Solid-State Circuits*, 23:750–758, June 1988.

[25] F. Lin, L. Liu, P. S. Kooi and M. S. Leong. "Design of MMIC LNA for 1.9 GHz CDMA Portable Communication". In *Proceedings of International Conference on Microwave and Millimeter Wave Technology*, pages 205–208, 1998.

[26] Fallesen C. and Asbeck P. "A 1 W 0.35 μm CMOS Power Amplifier for GSM-1800 with 45% PAE". In *IEEE International Solid-State Circuits Conference*, pages 158–9, 2001.

[27] Giry A., Fourniert J.-M. and Pons M. "A 1.9 GHz Low Voltage CMOS Power Amplifier for Medium Power RF Applications". In *IEEE Radio Frequency Integrated Circuits (RFIC) Symposium*, pages 121–4, 2000.

[28] H. A. Wheeler. "Simple Inductance Formulas For Radio Coils". In *IRE Proceedings*, pages 1398–1400, 1928.

[29] Ichiro Aoki, Scott D. Kee, David B.Rutledge and Ali Hajimiri. "Fully Integrated CMOS POwer Amplifier Design Using the Distributed Active-Transformaer Architecture". *IEEE Journal of Solid-State Circuits*, 37(3):371–383, March 2002.

[30] J. Lucek and R. Damen. "Designing an LNA for a CDMA front end". *Rural Electrification*, 22:20–30, February 1999.

[31] J. Zhou and D. Allstot. "A Fully Integrated CMOS 900 MHz LNA Utilizing Monolithic Transformers". In *IEEE International Solid-State Circuits Conference*, pages 132–133, 1998.

[32] K. L. Fong. "Dual-Band High-Linearity Variable-Gain Low-Noise Amplifiers for Wireless Applications". In *IEEE International Solid-State Circuits Conference*, pages 224–225, 1999.

[33] M. J. M. Pelgrom, A. C. J. Duinmaijer and A. P. G. Welvers. "Matching Properties of MOS Transistors". *IEEE Journal of Solid-State Circuits*, 24(5):1433–1440, October 1989.

[34] M. Koyama, T. Arai, H. Tanimoto and Y. Yoshida. "A 2.5-V Active Low-Pass Filter Using All npn Gilbert Cells with a 1-V_{p-p} Linear Input Range". *IEEE Journal of of Solid-State Circuits*, 28(12), 1993.

[35] P. Rodgers, M. Megahed, C. Page, J. Wu and D. Staab. "Silicon UTSi CMOS RFIC for CDMA Wireless Communications Systems". In *IEEE Radio Frequency Integrated Circuits Symposium*, pages 181–184, 1999.

[36] P. Wambacq and W. M. Sansen. *Distortion Analysis of Analog Integrated Circuits*. Kluwer Academic Publisher, Kluwer International Series in Engineering and Computer Science, 1998.

[37] Pradeep B. Khannur. "A CMOS Power Amplifier With Power Control and T/R Switch for 2.45-GHz Bluetooth/ISM Band Applications". In *IEEE Radio Frequency Integrated Circuits (RFIC) Symposium*, pages 145–148, 2003.

[38] Q. Huang, P. Orsatti and F. Piazza. "Broadband, 0.25 μm CMOS LNAs with Sub-2 dB NF for GSM Applications". In *IEEE Custom Integrated Circuits Conference*, pages 67–70, 1998.

[39] R. G. Meyer and W. D. Mack. "A 1-GHz BiCMOS RF Front-End IC". *IEEE Journal of Solid-State Circuits*, 29:350–355, March 1994.

[40] R. Moroney, K. Harrington, W. Struble, B. Khabbaz and M. Murphy. "A High Performance Switched-LNA IC for CDMA Handset Receiver Applications". In *IEEE Radio Frequency Integrated Circuits Symposium*, pages 43–46, 1998.

[41] Ravi Gupta, Brian M. Ballweber and David J. Allstot. "Design and Optimization of CMOS RF Power Amplifiers". *IEEE Journal of Solid-State Circuits*, 36(2):166–175, February 2001.

[42] Robert F. Pierret. *Semiconductor Device Fundamentals*. Addison-Wesley Publishing Company, Inc., second edition, 1996.

[43] Sowlati T. and Leenaerts D. "A 2.4 GHz 0.18 μm CMOS Self-biased Cascode Power Amplifier with 23 dBm Output Power". In *IEEE International Solid-State Circuits Conference*, pages 294–467, 2002.

[44] Stephen A. Maas. *Nonlinear Microwave Circuits*. Artech House, Inc., Norwood, MA, 1988.

[45] Su D., Zargari M., Yue P., Rabii S., Weber D., Kaczynski B., Mehta S., Singh, K., Mendis S., Wooley B. "A 5 GHz CMOS Transceiver for IEEE 802.11a Wireless LAN". In *IEEE International Solid-State Circuits Conference*, pages 92–449, 2002.

[46] Sunerarajan S. Mohan, Maria del Mar Hershenson, Stephen P. Boyd and Thomas H. Lee. "Simple Accurate Expressions for Planar Spiral Inductances". *IEEE Journal of Solid-State Circuits*, 34(10):1419–24, October 1999.

[47] Tajinder Manku. "Microwave CMOS-Device Physics and Design". *IEEE Journal of Solid-State Circuits*, 34:277–285, March 1999.

[48] Theerachet Soorapanth and Thomas H. Lee. "RF Linearity of Short-Channel MOSFETs". In *First International Workshop on Design of Mixed-Mode Integrated Circuits and Applications*, pages 81–84, July 1997.

[49] Thomas H. Lee. *The Design of CMOS Radio-Frequency Integrated Circuits*. Cambridge University Press, 1998.

[50] W. M. Snelgrove and A. Shoval. "A Balanced 0.9-um CMOS Transconductance-C Filter Tunable over the VHF Range". *IEEE Journal of of Solid-State Circuits*, 27:314–323, March 1992.

[51] Yongwang Ding and Ramesh Harjani. "A + 18dBm IIP3 LNA in 0.35μm CMOS". In *IEEE International Solid-State Circuits Conference*, pages 162–3, 2001.

[52] Yongwang Ding and Ramesh Harjani. "High Linearity Amplifiers in CMOS". In *Communication Design Conference*, 2003.

[53] Yongwang Ding and Ramesh Harjani. "A CMOS High Efficiency +22 dBm Linear Power Amplifier". In *IEEE Custom Integrated Circuits Conference*, 2004.

[54] Z. Wang and W. Guggenbuhl. "A Voltage-Controllable Linear MOS Transconductor Using Bias Offset Technique". *IEEE Journal of of Solid-State Circuits*, 25:315–317, February 1990.

About the Authors

Yongwang Ding received the Ph.D. degree and the M.S. degree in Electrical Engineering from University of Minnesota, in 2004 and 1999, respectively, and the B.S. degree in Physics from Peking University, China, in 1996. He joined Bermai, Inc. in May of 2001 and has been a RF IC design engineer since then.

Ramesh Harjani (S'87-M'89-SM'00) received the Ph.D. degree from Carnegie Mellon University, in 1989, the M.S. degree from the Indian Institute of Technology, New Delhi, in 1984, and the B.S. degree from the Birla Institute of Technology and Science, Pilani, in 1982 all in Electrical Engineering. He is currently an Associate Professor in the Department of Electrical and Computer Engineering and a Member of the graduate faculty of the Department of Biomedical Engineering at the University of Minnesota. Prior to joining the University of Minnesota, he was with Mentor Graphics Corp. in San Jose, California where he worked on CAD tools for analog synthesis and power electronics. He has had a number of short stints as a member of the technical staff at Lucent Technologies, Allentown and worked on data converters for DSL. He co-founded Bermai, Inc, a startup company developing CMOS chips for wireless LAN applications in 2001. His research interests include analog/RF circuits for wired and wireless circuits, low power analog design, sensor interface electronics and analog and mixed-signal circuit test.

Dr. Harjani received the National Science Foundation Research Initiation Award in 1991. He has been a pioneer in the analog circuit synthesis community and received a Best Paper Award at the 1987 IEEE/ACM Design Automation Conference and the 1989 International Conference on Computer-Aided Design. He received an Outstanding paper award at the 1998 GOMAC. His research group was the winner of the SRC Copper Design Challenge in 2000 and the winner of the SRC SiGe challenge in 2003. He is a co-author of the books, Design of Modulators for Oversampled Converters (New York: Kluwer, 1998), and Design of High-Performance CMOS Voltage-Controlled Oscillators, (New York: Kluwer, 2002). was an Associate Editor for IEEE Transactions on Circuits and Systems Part II, Analog and Digital Signal Processing from 1995 to 1997 and is currently a Guest Editor for the International Journal of High-Speed Electronics and Systems, Special issue on "High-Speed Mixed-Signal Integrated Circuits" 2004. He was the Chair of the IEEE Circuits and Systems Society technical committee on Analog Signal Processing from 1999 to 2000 and a Distinguished Lecturer of the IEEE Circuits and Systems Society for 2001-2002.

Index